PRAISE FOR *SOIL*

'A love letter to Mother Earth and entertaining must-read that goes to the heart of our survival.' **Charles Massy, author of *Call of the Reed Warbler***

'It's a huge task to get a city girl like me who kills pot plants to read thousands of words about soil, but Matthew Evans has done it. Page after page of revelation, making visible the complex and vital world beneath our feet. Reading this book is going to make you dig dirt.' **Dr Rebecca Huntley, author of *How to Talk About Climate Change***

'An exuberant, intelligent, mind-expanding hymn to the soil – sung from the heart of a man who has experienced its miracles. Wonderful stuff.' **Isabella Tree, author of *Wilding***

'This book is an urgent and passionate plea to take soil seriously, not just for farmers, gardeners and cooks, but for anyone who eats.' **Gabrielle Chan, author of *Rusted Off***

'Soil is everyone's business and this book pulls no punches. It lays bare our reliance upon the intricate life beneath our feet. A must-read in every school, local library, community garden, university and for your bookshelf.' **Costa Georgiadis, host of *Gardening Australia***

'A fascinating read about one of the most important issues facing our planet – the health of the earth, literally.' **Peter Gilmore, Quay**

'This book is for anyone who eats, and therefore benefits from the gifts that are bound up in this soil we all share. Whether you're a gardener, a farmer or just an enjoyer of food, the story of our soil is part of our humanity, and our future. Matthew Evans has done a brilliant job of inviting us into the mysteries, stories and understandings of this stuff right under our feet, largely disregarded, but which not a single human alive can do without.' **Kirsten Bradley, Milkwood Permaculture**

'There's no human health without plant health, no plant health without soil health, and no soil health without terrific books like this one. Matthew Evans has written a robust manifesto for the largest underground movement in the world.' **Damon Gameau, filmmaker**

'A real page-turner. Matthew Evans will take your understanding of soil from being "dirt" to being a precious and magical resource. With lessons in history, biology, sociology, politics and war, once you're done, you will defend any little or large patch under your care with your life, understanding its utmost importance to the survival of life on earth as we know it.' **Alexx Stuart, author of *Low Tox Life***

PRAISE FOR *ON EATING MEAT*

'Compelling, illuminating and often confronting, On Eating Meat is a brilliant blend of a gastronome's passion with forensic research into the sources of the meat we eat. Matthew Evans brings his unflinching honesty – and a farmer's hands-on experience – to the question of how to be an ethical carnivore.' **Hugh Mackay, author of *Australia Reimagined***

'Intellectually thrilling – a book that challenges both vegans and carnivores in the battle for a new ethics of eating. This book will leave you surprised, engrossed and sometimes shocked – whatever your food choices.' **Richard Glover, ABC Radio**

'Matthew Evans fearlessly investigates where our food comes from and the hidden impacts of our industrial food system. If you eat meat, read this book.' **Anton Vikstrom, Good Life Permaculture**

'Insightful, well-researched and highly readable, Matthew Evans' new book On Eating Meat presents an honest and challenging assessment of the livestock industry and the ethical and environmental issues surrounding our consumption of meat … this book equips the reader with the knowledge to get beyond the entrenched opinions of its topic area, and allows us to decide whether and what type of meat we wish to consume, and with what consequences for the future.' **Professor Andy Lowe, Director of Agrifood and Wine, University of Adelaide**

'This is the most important food book I've read in years. Not just for meat lovers or vegans, it should be read by anyone who eats food. When I finished this book I felt informed, connected and empowered to make better decisions about how I shop, cook and eat.' **Alex Elliott-Howery, Cornersmith**

Soil

Soil

The incredible story of what
keeps the earth, and us, healthy

Matthew Evans

Published in 2021 by Murdoch Books, an imprint of Allen & Unwin

Murdoch Books Australia
83 Alexander Street, Crows Nest NSW 2065
Phone: +61 (0)2 8425 0100
murdochbooks.com.au
info@murdochbooks.com.au

Murdoch Books UK
Ormond House, 26–27 Boswell Street, London WC1N 3JZ
Phone: +44 (0) 20 8785 5995
murdochbooks.co.uk
info@murdochbooks.co.uk

 A catalogue record for this
book is available from the
National Library of Australia

A catalogue record for this book is available from the British Library

ISBN 978 1 92235 141 8 Australia
ISBN 978 1 91166 819 0 UK

Cover and text design by Daniel New
Typeset by Midland Typesetters
Printed and bound by CPI Group (UK) Ltd, Croydon, CR04YY

14

The paper in this book is FSC® certified.
FSC® promotes environmentally responsible,
socially beneficial and economically viable
management of the world's forests.

To all the soil scientists out there, whose research never gets the attention it deserves. And to the gods of soil, whose gifts we need to cherish more.

Contents

Preface

It feels like I'm walking on air. Or a trampoline. The soil under my feet springs and bounces back, cushioning my steps, dampening any sound, lifting my spirits.

I'm in the Tarkine, the second-largest temperate rainforest on Earth, tucked up on Tasmania's rugged north-west coast. I'm wending my way through an ancient myrtle beech forest, inhaling air that revitalises my lungs and elevates my mood. It's doing me incomparable good. Some of the trees in this forest were around when Julius Caesar ruled in Rome. Over 90 species of moss, 150 species of liverwort and innumerable lichen smother the trees and the ground. Mushrooms, the fruiting bodies of fungi, sprout forth in vibrant gold, impossible bright blue and pinot red. There are countless fungi here not recognised, yet, by science. And they're just the ones big enough to see.

Beneath my toes lies humus, the dark, rich matter made from the decomposition of former trees and plants. There are also living arthropods, tiny little creatures with jointed limbs. If this forest is anything like ones in North America, each footfall is being supported by over 120,000 tiny legs on 16,000 little critters. But we wouldn't know for sure, because the soil in this forest, unlike that in Oregon where those numbers are known, is an unstudied wilderness.

Ten years ago, on this same walk, I would've been focused on the trees – some of the tallest flowering plants on Earth. I'd have looked up, to wonder which of over 100 species of bird had called. In the black, reflective rivers, I'd look for platypus, or the world's biggest freshwater crayfish, along with a few other unique burrowing crays. In this land, an ark for species not found anywhere else, there is plenty to get astonished about. And I still do get astonished. But these days I'm also entranced by what lies beneath. By soil. I imagine it breathing, as I know it must. The life within it, billions and billions of microbes under each of my boots, is inhaling and exhaling. The carbon released from soil is the main food source – carbon dioxide – for the trees above. They then feed the microbes who fed them; the original circular economy. I imagine the thousands of kilometres of worm tunnels that must crisscross under each hectare of this land. I try to envisage the myriad underground interactions as the plants communicate with each other, and with the whole ecosystem in the soil, as they tell tales of their lives, their age, their needs, their threats.

The fact that soil can help a tree stand tall is no mystery to me. Yet for the last decade I've gradually come to appreciate that what we see is only a reminder of a far bigger system, a far more complex social structure, which happens underground.

Beneath my feet, in the impossibly healthy soil of this untouched forest, lie trillions of living things from tens of thousands of species. Each one is involved in a life and death struggle. Finding allies, fighting foes. Each one feeds something else, including the trees that shade my walk. It's a hotbed of life, death and sex. Beneath my boots lies way more carbon than in the air above my head. The weight of all the bacteria on Earth, most of which live in soil, is 35 times greater than all the animals on the earth and in the seas. There's a lot more biodiversity *in* the earth than above it.

This forest, this landscape, is made by this soil. It's a far cry from the cracked and often bare earth I walked in my youth, around the nation's capital, Canberra. The Tarkine gets more rain, up to three metres a year – and everything above the ground is a reminder of the magic that happens under our feet. This forest has more things

I can see, but way more things I can't see, than the land where I first laid my feet.

Now, when I walk in the Tarkine, the visible part reminds me of the living nature of the soil underneath. To get to this point, to see soil as more than just dirt, took me a while. It took some relearning of simple science and a reframing of biology, some deep research and a leap of faith. And it just kept getting more fascinating the more I dug into it.

I'd like everybody to get as excited about soil as I am. I want everyone to shed the world above ground for a while, to see the world from the ground up, and to feel the astonishment of witnessing where the overwhelming majority of life on Earth starts and ends. Appreciating something so commonplace and yet miraculous that has been right under our noses the whole time is a great revelation.

For the next few pages I invite you to join me as I sing the praises of soil.

From the Ground Up

Growing food isn't rocket science. It's way more complex than that.

Remember 2015 and its relevance to soil? Surely you must. It was declared the International Year of Soils by the United Nations' Food and Agriculture Organization. Don't remember 2015 for that reason? Well, neither do I.

What did you do on 5 December last year, then? Celebrate World Soil Day? I didn't think so.

Let's face it, soil has an image problem.

It doesn't speak. It doesn't move of its own accord. It doesn't give us visceral pleasure like a forest or a waterfall. The most wonderfully complex and the most important part of Earth involved in feeding us, healing us, nurturing us and sustaining us isn't breathtakingly pretty. It isn't poetic. It lies just beneath our feet, but well out of our minds and collective consciousness. It's the part of Earth we've long overlooked. It's the stuff we've wiped our collective feet on for a couple of hundred thousand years, and often – to our detriment – ignored.

Soil, however, is at last beginning to be truly understood, and its role rightly esteemed. The threads of science are tying together the story of this amazing and underrated resource. At this pivotal time, with a climate collapse looming, with human ingenuity put towards growing the least nutrient-dense foods in history, while simultaneously causing alarmingly high levels of environmental damage, we are at a crossroads. What happens now – in human and global health – depends on how we look at the land beneath our feet.

At long last, we are beginning to understand the importance of soil.

I didn't come to this conclusion easily. My youthful imagination wasn't fired by the 'chocolate cake' texture of good soil that has long excited gardeners and soil scientists. Until relatively recently, I found more joy and excitement in human culture, in David Attenborough's explorations of the wild world, and in *actual* chocolate cake, than I did in thinking about dirt as anything more than something that accumulated underneath my nails.

For me, soil seemed dull and insipid. A dry, dusty topic. A medium for growing things. Something to hold up trees and stop my house descending into the Earth's core. Isn't soil just crushed-up rocks and minerals, a lifeless mélange of sand and clay where plant roots go to find water and nutrients? Don't we simply grow food where soil is good, and don't bother where soil is bad? Aren't we just the beneficiaries of the geological lottery of where soil has naturally formed? Can't we, in modern times, just throw fertiliser on pretty much any piece of land, no matter how poor, and add seeds and water and get some kind of crop?

Well, yes and no. The truth about soil will astound and amaze you. It will blow your mind. I've been digging into the area for years, constantly delighted and surprised by what I've found.

But before we start to unpick what's so great about soil and how it affects our own lives, it's worth at least acknowledging why we have neglected the land that feeds us. It's hardly surprising that soil gets overlooked. The narrowly defined area of soil science is populated by a certain breed of people who talk in impenetrable words about impenetrable topics. This is very normal in the scientific community, and that's just fine. Let the researchers research.

For a long time, we've often looked down on those who work the soil. How many times have I heard 'they're just a farmer', in everything from fairytales to common conversation? The majesty of growing food has been masked by a culture that doesn't value or understand the incredible scope and nuance possible in farming and gardening.

Modern Western culture also tends to look at serious home gardeners as quaint. Or idealistic. Self-satisfied, homespun and essentially dull people who would rather wear hand knits than go to a rave. Growing food suffers from the 'dead boring' moniker, when it really should be revered. The ability to turn sunlight into nutritious food has literally fuelled civilisations, built societies, and allowed us to expand the human population exponentially. That you should ever hear the term 'just a farmer' would be exasperating if it weren't actually ignorant and demeaning.

That's the world we live in, though. Now, after a good twelve years of growing food in a small patch of southern Tasmania, in the southernmost shire in Australia, having watched as I turned good soil bad, and the other way around, I find the scales falling from my eyes like cracked mud flaking from my boots. A confluence of ancient wisdom and recent discoveries has led me to revere the earth. And what I want to shout from the mountaintops is the notion that a love of land isn't just for diehards. What happens in the fine margin of earth sandwiched beneath our feet and the rocks below has implications for all of us – our communities, our climate, our health and for the globe as a whole.

Soil nourishes us in more than just the obvious ways. Good soil isn't just an abstract concept; it's a thing of wonder. Far from being an inert, dull, stable entity, it's teeming with life, and forever in a state of flux. On the visible scale, there's little to hint at the wondrous activity and industry that lies within.

There are more living things in a teaspoon of healthy soil than there are humans on Earth. Soil hosts a startling array of life. Despite being relatively unstudied, about a quarter of all living things that have been identified, anywhere, live in soil. The majority of wild bees nest in soil, for one thing – a full 14,000 species[1] – as do other pollinators such as native wasps. Soil houses ants and spiders and termites, in numbers – and mass – that make above-ground creatures look sparse.

But wait, it gets more amazing. There's an interconnectedness that means that soil is key to life on land. Living, healthy soil is pivotal in good ecosystems. It allows plants to communicate with each other. Over 70 per cent of all multicellular animals on Earth live in soil,[2] and it's only through the microbes in soil that plants can thrive. Most of our antibiotics are derived from soil microbes. Good soil holds and sheds water better than poorer soils, and stores way more atmospheric carbon. And as we'll see later, the billions of microbes in each teaspoon of healthy soil can form the genesis of rain.

As a field of study, however, soil hasn't been getting the kudos it deserves. For too long, rather than exploring what's right under our feet, we have looked to the stars, to sexier challenges like flying to the Moon, or exploring the oceans deep. For aeons we've looked at the growing of crops through either a lens of woo-woo – the mystical and abstract – or the reductive, where synthetic fertilisers and yield are king.

Yet now, with complex DNA analyses and intricate scientific studies overturning abominable results in the field from an ill-informed or simplified understanding of soil biology, we can see a truer picture.

What emerges is an astonishing, resilient life force. A complex ecosystem that forms part of the magnificent whole, where life begets life, and soil confers on the globe more than just something to stand on, or dig up. Healthy soil is what we need to be happy, healthy humans, and to regenerate landscapes. If we continue down the path we've followed since humans started cultivating land, we'll ruin the earth that feeds us, growing nutrient-poor food in an unstable climate that threatens to spiral out of control.

We have the capacity to slow climate change, grow an abundance of better-tasting, nutrient-dense food, gradually alter the very genetic makeup of our bodies for the better, *and* help heal the world. The answer is right in front of us and all around and underneath us.

All we need to do is stop treating soil like dirt.

CHAPTER 1

What You Eat is Made out of Thin Air (and a Tiny Bit of Dirt)

When I was a kid living in the suburbs, my parents kept indoor plants. One in particular, a fiddle-leaf philodendron, did something very odd. It was a single-stemmed vine that shot up from its pot behind the telly. It grew so tall that we'd pin it up to the wall, then later against the ceiling, as it climbed. Over the years it rose right over the huge windows at the front of the house and across to the other wall, spanning the entire width of our lounge room.

We never fed this philodendron. We'd tip a little water into its terra-cotta pot every few days, which eventually rotted the carpet underneath. But we never added anything to the soil. And the soil level didn't drop. As each year brought increasing growth, the amount of plant material clinging to the walls and ceiling eventually outstripped the amount of soil in the pot. It seemed odd, because all the solid plant matter didn't come from the earth in which the vine grew, but from somewhere else. To be honest, it didn't engage my brain too much. It just seemed bizarre and left a question in my mind that took decades to answer.

This book is about soil. Except it's not. It's about soil *and* all the things soil does for us, including how it fuels our bodies. Even more important is how our land provides us with thousands of nutrients that help keep us healthy.

To understand that, we have to get right back to basics. We need to understand how plants actually grow, and hence how our food is produced. It's pretty magical, and to this day blows my mind.

Put simply, plants do something no animal can do: they make their own food.

The same puzzle that I saw in my childhood pot plant confronted Jan Baptista van Helmont in the early 1600s. Here, in the hands of an accomplished intellectual, it wasn't just a riddle, it became an important scientific discovery.

Van Helmont was a Flemish physician and scholar who rejected the idea that there were only four elements – earth, fire, water and air. He believed in rigorously testing his theories. So, in a simple experiment with far-reaching consequences, van Helmont decided to investigate the growth of plants in a long-term study that was the first of its kind, and became his most famed experiment, despite some other incredible work in his lifetime. He planted a 2.2 kilogram (5 pound) willow in 90 kilograms (200 pounds) of soil and kept it indoors for five years, adding a measured amount of water as needed. After the five years, the tree was weighed again. From a paltry start at just over 2 kilograms, the tree came in at 77 kilograms (170 pounds), and the soil had lost only 57 grams (2 ounces). ONLY 57 grams!

So where, oh where, had the extra 75 kilograms come from? The only input van Helmont could discern was water, so he reasoned it was the water that somehow became the plant matter. As a concept, I still find it hard to understand. How can something solid grow out of seemingly nothing? As far as we know, this was the first time anybody proved without doubt that plants don't consume soil to fuel most of their growth.

Until van Helmont's time, gases were an unknown concept. (Incidentally, van Helmont was the first to identify carbon dioxide, and invented the word 'gas' to describe previously unknown components of air.) About a century and a half later, in 1782, Swiss botanist Jean Senebier showed that, in the presence of sunlight, plants take in carbon dioxide and release oxygen. In other words, that the miracle of photosynthesis exists.

The bit about the plant taking in carbon dioxide and releasing oxygen was the bit I did manage to absorb from lessons on photosynthesis at school. Plants make oxygen for us to breathe, was the lesson – but the bit about where the carbon goes next is the truly remarkable part.

Thanks to scientists like van Helmont and Senebier, and so many who followed, we now know that just about all energy on Earth comes from the sun. Even the heat buried down deep in the Earth is a result of the formation of the planet and solar system billions of years ago.

When I talk of energy on Earth that we use coming from the sun, I really do mean *all* of it, not just solar energy. Wave energy, wind energy, oil, natural gas, coal. All of these are only possible because of the effects of the sun.

Sunlight does many things. Some of its energy is trapped on Earth in the form of heat for a time – a very short time in geological terms. This trapped heat causes weather, which in turn causes wind and waves – energy we can trap and use for power.

As well as heat, the sun also gifts us light. And plants (and some other organisms) are able to trap the energy of that light and convert it into sugars, and then store that energy. This is photosynthesis – the topic your biology teacher droned on about, and Senebier first postulated.

But what photosynthesis really does is nothing short of wondrous.

When a plant takes in carbon dioxide – which, at an elemental level, is one carbon atom attached to two oxygen atoms – and releases oxygen, the carbon remains in the plant. The carbon that the plant doesn't exhale (the proper term is *respire*) is stored in the structure of the plant itself.

Consider this: carbon is the basis for practically all life on Earth. It's like the original building block for everything from the smallest bacteria to the largest mammal. Everything that is living is made up of carbon, at least in part.

Now, carbon alone isn't enough. Humans and other animals need something else. We need a source of energy, too – and that energy is also made of carbon. The thing that fuels us, and all other animals, is sugar, which in turn is a mix of carbon, oxygen and hydrogen – all common elements on Earth, and in the air. Sugar is the starting point for pretty much all energy that living cells use as fuel.

Sugar isn't confined to just one type of molecule, and can be found in complex forms, not simply as the white stuff in the kitchen pantry.

And where does this sugar – and hence all those fats and proteins that are made by other living things – come from?

Photosynthesis. Yes, high school science. But just how does that work again?

Photosynthesis is one of the most extraordinary things, and yet is so ubiquitous we simply take it for granted. Using sunlight as its energy source, a plant converts water and carbon dioxide into carbohydrate – sugar. Plants make sugar out of thin air. They make their own food! How insane is that? You can't do it. I can't. We humans can stand in the sun as long as we like, and all we get is a melanoma and hungry.

A plant can make the sugars on which pretty much all other life on Earth depends.

Not only that, those plants that are making our food are also making oxygen for us to breathe, as a by-product of the process!

The sciencey equation looks like this, where C is carbon, O is oxygen, and H is hydrogen:

$$\text{Carbon dioxide} + \text{water} = \text{oxygen and sugar}$$
$$6CO_2 + 6H_2O = 6O_2 + C_6H_{12}O_6$$

Now, here we will need to delve into some simple chemistry. Think of the elements as building blocks – Lego, if you like – where each element can link to others, and the combinations can be almost limitless, but each element only has so many links. All life on Earth is built from these blocks. That's the practical part.

But there's also the imaginative, magical part of Lego, where you don't just stack things, and where the transformation of simple blocks can create rounded shapes and teetering towers of unspeakable beauty. In the plant world, the carbon acts as the building blocks, harvested out of thin air, which can create a magnificent oak tree for us to swing under and picnic beneath. The carbon building blocks – with some oxygen, nitrogen and hydrogen – create the magnificent ancient rainforests of Tasmania's south-west. They form the alpine meadows of central Italy, the verdant grasslands of America's Midwest, the rhododendron forests of eastern Nepal. Plants use their sugars to build themselves into myriad forms on every habitable landscape.

To my mind, the making of sugar out of air (and water) is sheer alchemy. A miracle. Without these sugars, no animals on Earth could exist – only other things that can photosynthesise, and a handful of specialised bacteria that can feed off sulphur, methane or the heat from lava.

Without sugar, there'd be no fungi. We'd not have most species of bacteria. There'd be no elk, no orangutans, no platypus, no ferrets. No animal life as we know it. The vast, vast majority of life on land is dependent on plants. If plants didn't make sugars, there'd be little else.

These sugars made by plants become cellulose, and are synthesised into lignin, the polymer that forms the hard trunks of trees; wood. The sugars give plants energy to help fight disease, or bad weather, and to repair themselves after being nibbled by a passing giraffe.

These sugars also feed the rest of the world, because the rest of world can't create its own food out of thin air. Plants provide the most important ingredient of all – energy in digestible form.

So, a plant can create its own food, and our food. But of course, plants aren't just made of sugars. They also contain other things.

9

In nutritional terms, these are expressed as macronutrients and micronutrients. In biological terms, they form the structure of the plant, and aid its metabolism.

Macronutrients, as the name suggests, are major nutrients, available in bigger quantities. Micronutrients are present in small, sometimes minuscule, amounts.

For us humans, there are only three macronutrients: carbohydrates (a fancy name for sugars), proteins and fats. Micronutrients include vitamins and minerals, and even smaller elements such as phytochemicals and antioxidants, many of which science hasn't yet properly identified, let alone deciphered precise roles for.

Plants, like us, need micronutrients and macronutrients to live. A plant's macronutrients are different to ours and include nitrogen, phosphorus, potassium, magnesium, sulphur and calcium, along with carbon, oxygen and hydrogen, which we met earlier.

Plant cells are mostly made of cellulose, a complex sugar arrangement. Cellulose isn't digestible by humans – but it does, however, provide vital fibre in our diet. Plants also make other sugars besides cellulose.

Plants also take up another gas from the air – nitrogen – which allows the production of protein.

We know, from eating avocado and using olive oil, that plants also make fats (which are also derived from carbon, oxygen and some hydrogen).

So, plants provide all of the macronutrients humans need – carbs, proteins and fats – from gases transformed and solidified from the air and water. But this is only possible because of the original trapping of energy from sunlight, using that sunlight to force apart water and carbon dioxide and create carbohydrates.

Human micronutrients are multiple, and there's no consensus on exactly how many are essential. Micronutrients might have calcium in their structure, or manganese, or iodine. So where do all of these other micronutrients come from?

The answer is soil – and they're made available to plants through some mighty complex biological processes.

The 57 grams that had vanished from van Helmont's pot of soil over five years, as his tree grew, is the source of food's micronutrients – the teeny-weeny amounts of trace elements such as boron, chlorine, copper, iron, manganese, molybdenum, zinc and nickel, and maybe iodine, cobalt and selenium, bound up in myriad chemicals in the plant.

What we do know about food, in general, is that humans have evolved to eat it. (Bet you're glad you invested in a book to tell you that.) But you wouldn't know that if you got all your information from food manufacturers, where they brag about the thiamine or other 'essential' nutrients they've put back into their products. Or if you got a brochure from your local medical centre sponsored by a dairy board or meat and livestock association.

Nutritional science is a relatively new and often overly simplistic field littered with bad advice or a lack of nuance. It's full of things that we are told to do, and not do, and seems to want us to believe in its cleverness. That the manufacturer has 'made' the food instead of just processing (and usually simplifying) it. I think if you want to *really* understand nutrition, then you really need to look at the big picture, not at what some industry lobby group wants to sell you.

For all of human history, up until about 10,000 years ago, all of us foraged and hunted for our food. Some still do a bit. Agriculture, of the style that you and I recognise as modern farming, began when humans stopped roaming and started to settle, domesticating plants and animals that were conducive to the task. That's a short phase in our long evolutionary past. Prior to that, we ate what we had at hand, or could chase down. According to the Food Solution Project, over 30,000 species of edible plants have been identified around the globe.[1] That's an awfully diverse range of plants, which indigenous cultures took full advantage of. It is estimated that up to 5000 species of edible native plants and animals were available to Australian Aboriginal people across the entire continent.[2]

Variety was key. And variety was the norm. As, apparently, was the occasional feast/famine cycle. Settling down in one place and growing food meant we could, in theory, control our harvests better, and store food for leaner times. This, many argue, allowed for the construction of bigger communities and population centres, more feasting, and an increasing non-labouring class who didn't actually do any growing or animal husbandry themselves. For a while, a longish while, we had farmers, priests and tax collectors, with a few aristocrats too posh to get their hands dirty thrown into the mix.

The upshot of this revolution in the way we sourced food, from foraging to farming, had profound ramifications on our diet. Over the centuries, we started to settle down with fewer and fewer of those hundreds or thousands of different species, concentrating on those we could grow with most success. So the animals we consumed were whittled down to what we have now, which is only five major land species – cattle, sheep, pigs, goats and chickens. And plants narrowed in scope quite rapidly, too.

For a long while, however, we didn't miss the multiplicity of foods a forager's diet boasted, as our diets were still supplemented with wild-caught and foraged food, and plant varieties were still relatively abundant. At one point it's estimated there were 400,000 different varieties of rice, for instance, all from a single species.[3]

Now, it's important early on to define the difference between a species and a cultivar or breed. A plant species could be, say, tomatoes. Then a cultivar (variety) is a tomato of a certain type – perhaps a black Russian, or green zebra. An animal species could be, say, a pig; a breed could be Large Black or Wessex Saddleback. (It seems this isn't necessarily common knowledge. I once spoke with a now former Australian federal agriculture minister, Barnaby Joyce, who didn't realise that fish of different names are actually different *species*, which isn't the same as *breeds* of animals like cattle, for example Hereford versus Angus.)

Diversity is really important. We used to eat a much more diverse range of plant and animal species over the seasons and the years. Today, in the West at least, only about 30 plant species provide about

90 per cent of our calories.[4] But also, we've decreased diversity within the crops we do grow. Narrow genetics, through selective breeding over the centuries, has produced far more valuable, efficient and potentially flavoursome crops. But that uniformity also means a lack of diversity. Think of the difference in nature between trees of the same species, such as one of the eucalypts like snow gum, versus the similarity we see in a modern crop of broccoli.

Nature fights for diversity. It's the very definition of evolution, to create diversity in response to attacks by predators, disease or unprecedented climatic events. Humans, on the other hand, have striven for consistency. Why does this matter? Because we aren't designed for simple diets with simple nutrients. Our bodies are made for the diversity we enjoyed for aeons, since the very earliest years of *Homo sapiens* history.

Our guts and bodies are made for the elusive 57 grams of elements that were drawn from soil into van Helmont's plant. But the more we understand about food, about what goes into our food, the more we realise that we aren't getting diversity. Not in the species we eat. Not in the varieties of plants and breeds of animals we eat. And not in the complexity of what those plants and animals contribute in the way of micronutrients, as has only recently been discovered.

The reason for so much of this most recent loss, it's now becoming apparent, all comes back to soil.

Interestingly, even in what we might think of as lean and hard times, we grew a greater variety of plants than we do now. In *The Last Food of England*,[5] food writer Marwood Yeatman describes how the number of vegetables in cultivation in the United Kingdom peaked at 120 species around the year 1500, and slumped to its lowest point in the 1970s, from where it has only partially rebounded. In other words, with no tractors, no artificial fertilisers, and no Ottolenghi cookbooks on the shelves, the Brits grew a greater variety of different species as peasant workers tending vegetables than they do as cashed-up urbanites with international palates.

Despite the British fondness for 'allotments' – those community-based vegetable gardens geographically removed from densely populated neighbourhoods – modern Brits don't eat variety. And the average Brit only consumes 1½ portions of the recommended '5 portions a day' of fresh produce. That works out to be 128 grams a day, down from 400 grams in the 1960s.[6] And while official statistics have 28 per cent of the population eating the recommended quantity of fruit and vegetables,[7] because these data include processed fruit and vegetables (frozen, canned and juiced products), this means tinned baked beans and apple juice made from imported concentrate are counted as 'fruit and vegetables'.

And it's not just the Brits. An extreme example, as is often the case, is the United States. There, the numbers are staggering: less than 1 per cent of adolescents, about 2 per cent of men, and only 3.5 per cent of women meet the national guidelines for consumption of fruits and vegetables. And that's just *volume*, not diversity.[8] More than half the average American's energy intake comes from what is known as 'ultra-processed' food[9] – something unrecognisable from the original ingredients, and produced in a factory.

In Australia, we're not much better. According to the Bureau of Statistics' 2018 health report, only 5 per cent of adults had a sufficient intake of both fruit and vegetables, with women scoring 8 per cent, and men 3 per cent.[10]

In developed nations, especially English-speaking nations, we simply don't eat enough greens.

Why do we eat so few plant varieties, and so little of them, when we're built for variety – and all the advice, from all the relevant authorities, has pretty consistently stated that we need a higher proportion of them in our diets?

One reason is because we've managed to make them taste bland. Modern growing, the type we've excelled in for the last hundred years, has stripped much of the flavour from food. The reason your child doesn't eat that carrot in their lunchbox comes down just as much to the carrot as your child's attitude. A fresh carrot, grown in healthy soil, actually tastes far better than one grown in poor soil. It's much easier to eat vegetables and fruits that are packed with flavour than ones that are insipid, woody or bland.

This truth goes to the heart of our food choices. It's actually the reason I discovered farming, and gave up life as a restaurant critic. Led by my gluttonous palate to find better food, I ended up in people's backyards, forsaking my job eating at the finest restaurants in the land. In an effort to feed myself better, I ended up breaking the ground and planting seeds myself. I once wrote a book, *The Real Food Companion*,[11] about what makes some foods taste better than others, even identical species being grown in the same geographical area.

What I discovered is that the almost intangible quality of the ingredients that my palate could taste represented the soul of the grower. And the soul of the grower is expressed in their treatment of soil.

CHAPTER 2

Soil, the Earth's Miracle Skin

Talking about soil is like dancing about architecture.
(with apologies to Elvis Costello)

Niels Olsen grabs a spade and strides purposefully into his front paddock. I'd been greeted at his house with a farmer's usual vice-like grip, and Niels' broad smile, before we stood and talked about life underground. The designer of a mulching, seeding, soil-building technique and machine, Niels is a font of knowledge on the topic of what makes good soil. But he wants us to get our hands dirty. Banging on about the earth is nothing compared to putting a spade in the ground and seeing what comes up. Not just seeing, but feeling and smelling. I reckon, left to his own devices, Niels would also be tasting soil, such is his fascination with the stuff that most of us take for granted.

What Niels was in a hurry to show me was 'chocolate cake'.

Ever since I started writing about food and farming, I've heard the expression that really good soil should be like chocolate cake. As a food writer and chef, I always thought that's fine, if you don't know the true nature of great chocolate cake. Sure, good soil can be brown like chocolate cake. And it can be a bit crumbly, I guess. But for a long while I thought if anyone actually believed good soil is like chocolate cake, they really needed to stop taking themselves so seriously. Or get an eye check. Or simply change the analogy.

And how wrong I was.

Niels' soil is the first true expression of chocolate cake I've seen in farmland. Like a good cake, it holds together well until you crumble it. It smells sweet. Unlike cake, it is also rich with visible life, both of the fungal and springtail type (springtails are like soil mites). You can also see worms. Lots of worms.

When I met Niels, it wasn't like I hadn't seen or touched soil. For nine years my partner Sadie and I had been managing our on-farm, organically run market garden. Our entire focus and stated aim was, and is, to 'farm soil'. But still, after nearly a decade of care, with the addition of compost and seaweed, the soil in our no-dig, no-spray market garden isn't yet like chocolate cake. It's still dustier in summer, claggier in winter, repelling or holding water in ways that truly great soil doesn't.

Chocolate cake soil has something called 'tilth'.

Tilth is a wonderfully descriptive word for the texture of soil that all growers strive for. It crumbles gently through your fingers when loosened, and clumps gently together (not stickily like clay) when pressed. Tilth demonstrates the capacity of soil to absorb and retain moisture in a way that doesn't cause waterlogging.

To understand tilth, it's good to get your head around the structure of soil – and for that, we need to venture a long way back in time.

As we all know, the Earth is mind-bogglingly old. In this book I'm going to delve into some pretty big topics that cover some pretty big numbers. Numbers so huge that the human mind isn't always very adept at comprehending them. But without referencing the numbers, it's really hard to appreciate how astonishing soil is, and how easily we can squander what little we have of it.

So, the Earth is about 4.6 billion years old. That's 4,600,000,000 years old. The number looks bigger, but more real, when you put all the zeros on it. That's about 57,500,000 human lifetimes, at a current life expectancy of around 80 years. Well over 50 million human lifetimes. Why am I going on about the age of the Earth? Because to understand soil, we need to look at how it was formed, and why a human life is a poor time frame in which to judge our actions.

Soil isn't just dirt. It's dirt with a few extra bits. It can be divided into the physical (rocks and dead organic matter), the biological (the living part), and the chemical – the interplay of elements from both the physical and biological parts with plant roots.

Soil, in the most basic sense, is a mix of rock crystals, air and water, along with life, and what's called organic matter – essentially dead stuff. We'll delve more deeply into the life of the soil later. Suffice to say here, soil life is pretty important.

Approximately 95 per cent of soil mass is made of sand, silt and clay – in other words, broken-down rock. This rock helps shape the physical structure of soil, the soil architecture that helps house life.

This is why we're going a long way back.

After the Earth was formed – a great, seething super-hot mass of gases and elements pulled together by gravity 4.6 billion years ago – it took about another half a billion years for it to reach any kind of solid state, as a crust started forming. All the elements – the iron, the calcium, the zinc that we want in soil today – were already pretty much present at that stage.

Eventually, the Earth's surface started to cool, and water was produced. We gained oceans about 3.8 billion years ago, by current reckonings. As the gases that formed our planet cooled below boiling point, water vapour started condensing. So it rained, probably for centuries. CENTURIES! And you think living in rainy old England (or Tassie in winter) is bad!

About a billion years after forming – so, 3.6 billion years ago, and only a few hundred million years after the formation of oceans – we think the first life on Earth began. Theories abound as to how and why. It's complicated, so let's just take it as read: Life began. These were single-celled organisms, bacteria mostly, that had the place pretty much to themselves for another 3 billion years before multicellular life formed. Bacteria and other single-celled organisms, such as archaea (which are very similar to bacteria), ruled. These bacteria could feed on sulphur, perhaps, and later on carbon dioxide and sunlight.

Pretty much at the start of life on Earth, the first cyanobacteria were formed. These were early bacteria that photosynthesised and didn't need oxygen. It's thanks to the actions of cyanobacteria, converting sunlight, carbon dioxide and water into oxygen and sugars – as we saw with plants in the last chapter – that we started to see atmospheric oxygen being produced. In fact, pretty much all the oxygen in the Earth's atmosphere is due to the work of cyanobacteria. Some oxygen is also formed by free-floating single-celled organisms in the ocean, known as phytoplankton – and sometimes it's as a guest of a plant cell.

Amazingly, it's the cyanobacteria that live as guests *inside* plant cells, within the leaves, that actually do the photosynthesising for modern plants.

Oxygen is the third most common element in the universe, so it's probably not surprising that we have it in the atmosphere. But it took a while for any oxygen released by the cyanobacteria to be trapped close to the Earth; scientists estimate complete oxygenation took about another billion years. And it's only then, plus another billion years, that conditions were ripe for much other life. (This other billion years has been dubbed by scientists the 'boring billion'. See, even scientists can find this stuff impenetrable …)

The presence of atmospheric oxygen put several conditions in place ready for the creation of soil. Oxygen in the air meant that rocks broke down in new ways, thanks to a newly arrived chemical reaction: oxidation. A thing called the Great Oxidation Event doubled the number of possible minerals in soil. It helped detoxify arsenic, and rapidly changed the availability of other minerals, such as sulphur and iron. It helped remove methane from the atmosphere. In other words, it rapidly escalated the possible diet of new life.

So we have bacteria and archaea, which as we will see later are vital for soil. We also have a whole heap of new components for soil, and an atmosphere that is suited to other forms of life. But before we could get true soil, another big change had to take place to crush rock far quicker than the movement of tectonic plates ever could.

Scientists now think there was one stonkingly big freezing event that gave sudden rise to more complex life on Earth. German and Australian scientists have coined the pivotal phenomenon Snowball Earth, which occurred about 700 million years ago, when ice covered virtually the entire globe. This was during the appropriately named Cryogenian period (with *cryo* meaning cold). And it was very, *very* cold. The Earth probably looked like an enormous white golf ball, it was so smothered in ice.

The massive freezing and thawing of Snowball Earth provided the necessary shift in the Earth's structure for multicellular life to really take off. Colossal glaciers did the work that rain and tectonic shifts couldn't. These glaciers ground down entire mountains to dust (or more accurately, sand, silt and clay) – which also released trapped nutrients such as iron, zinc, phosphates and potassium. This was followed by a massive deglaciation, where the ice sheets melted and retreated, which fractured whole mountain ranges and washed broken rock into the lowlands, valleys and oceans.

So, suddenly, as the land warmed again, the climate was more favourable to multicellular life. The atmosphere was oxygenated, and there were the building blocks of soil, as well as copious amounts of nutrients available. The age of bacteria certainly didn't end, but it did dovetail nicely into the age of complex plants – and once the plants arrived, they could trap the sun's energy and create way more sugars, which then formed the basis of life for animals.

While we haven't seen global freezing events comparable to the Cryogenian for the last 635 million years, we have had ice ages, and we still see the same processes at work. More often, these days, it's the slow erosion of rock that gives birth to soil. The act of freezing and thawing cracks rocks. Wind and rain gradually loosen rock crystals, which end up mounding against hills or at the bottom of valleys.

Where more recent glaciers have carved the Earth, such as during the last ice age about 12,000 years ago, soil is usually more abundant.

When we think of the Earth, we often think of it as the bit we see. The mountains. The plains. The rivers. The land that grows the forests. But all of those forces, and all of those elements that originally formed the planet, are still pressing inwards, and the Earth is actually still partly molten, way beneath our feet. If we could tunnel through the planet, and look out from a glass elevator, we'd be able to see all the things that don't look like the planet we call home, but make it what it is. On the descent, first we'd pass a veneer of topsoil, then we'd see dirt and clay, and then rock. After the rock, we'd pass into magma – molten rock – and then eventually into the Earth's iron core.

Why am I telling you this? Because it shows that the mineral component of soil isn't in short supply, despite some localised natural and man-made depletion on the planet's surface. We have no shortage of the raw ingredients that make up 95 per cent of the Earth's soil.

Functional soil also has organic matter in it – carbon-based molecules made out of formerly living things. It also has living microbes. Taken together, these physical and biological components need one other thing to make the chemical part work. It needs to have living plants growing in it – or to have had them recently growing in it.

Soil without life, including plant life, is dirt.

Now, if this soil is so important, how much is there? Well, it sounds like a lot if you say that about 100 million square kilometres (38 million square miles) of Earth is covered in topsoil.[1] But topsoil is unevenly spread over that land, and even at its deepest is only wafer thin, relatively speaking. It's a tiny fraction of the Earth, despite its vital role in our lives, covering just a bit over a quarter of all the land on Earth – but only about 7.5 per cent of the Earth's surface is available as potential agricultural land to grow our food.[2] The rest is ice, water, stone, too dry, too wet, too hot, or covered with cities.

Where there is no soil, there is limited life on land. Very few things of any size can exist in our terrestrial environment without soil.

Then, let's consider the Earth's depth. Think of the Earth as it can be seen from the Moon: as a fairly round ball. To reach its innermost centre, you'd have to travel over 6300 kilometres (4000 miles) to the Earth's core. If it took a year to travel through to the centre of the Earth, a feat that humans have not even come close to doing, how long do you think we'd be in the solid crust, which makes up the building blocks of most of our soil? Less than two days.

The Earth's rock crust is 30–45 kilometres (18–30 miles) thick, depending on where you're standing. The depth of the topsoil, however, is about 15 centimetres (6 inches), on average. So, if it took a year to reach the centre of the Earth, you'd spend about 0.74 of a second of the first day in the topsoil. That's right, less than a second of the entire year.

If you think of the Earth as our mother, then the topsoil isn't anywhere near as thick as her skin. It's thinner than the first layer of her epidermis. If Mother Earth was a person, the topsoil would be thinner than a single cell on the outside of her body.

Despite its thin existence, soil is the most biodiverse environment on the planet. In a single shovelful of healthy soil, according to the University of Illinois, there are more species than can be found above ground in the entire Amazon rainforest.[3] And the Amazon rainforest is the most biodiverse ecosystem on land.

But this thin veneer that gifts us life is in peril. David Pimentel, the late American ecological researcher from Cornell University, estimated that about 40 per cent of the world's agricultural land had been abandoned because it was no longer fertile.[4]

Even the land we haven't abandoned is in trouble. By some estimates, including those of Maria Helena Semedo, Deputy Director-General of Natural Resources in the United Nations' Food and Agriculture Organization, the FAO, we only have about 60 years of topsoil left if we keep growing food, and clearing land, using the same old methods.[5] If we keep buggering up topsoil in the same way as we have over the last 50 years, where we lose a soccer pitch of soil every five seconds through erosion or desertification.[6]

The FAO estimates that 40–50 per cent of the world's current agricultural land is degraded or seriously degraded, affecting up to 3.2 billion people.[7] The part of the earth that does all the growing – the top 10–15 centimetres – is over-utilised, underappreciated, and under pressure. If we want to work out better ways of doing things – how to conserve, preserve and perhaps improve our soils – then we have to understand what we actually have.

We have to start to look at things from the ground up, not the other way around. This applies equally to your backyard as it does to the world's biggest cattle station.

Great soil is made of a thing called loam.

Loam, by definition, is about equal quantities of sand, silt and clay. Too much sand, and nutrients and water can pass through too quickly. Too much clay, and the texture of the soil is pasty, easily saturated with water, and can be lacking in air. Loam is the best base, but good soil can be built with varying success at different ratios, too. (See the jar test at the end of this chapter to test your own soil.)

These basic rock crystals – sand, silt and clay – aren't in equal supply everywhere. Nature's sweepstake has endowed some places with way more of one than another, or very little of any. These rock crystals are inevitably the victims of gravity: they gradually move downhill, into creeks, rivers, and eventually the deep ocean. Luckily, the planet does keep making more of them. Freezing and thawing, storms, lightning, glaciers, earthquakes and the slow drip of water all help break down stone and rock. Soil life can do it, too.

While all those rock crystals provide some structure to soil, they're more like bricks in a house: they don't really stay together in the way we want, unless there is some kind of glue or mortar. A lot of the structure and mortar of soil, and a lot of the workings of soil, come from the life within it. The only way to get the holy grail of soil textures, the 'chocolate cake' of soil, is through abundant subterranean life. As Jon Stika, author of *A Soil Owner's Handbook*, says about soil life, 'Without biology, soil is simply geology.'[8]

In our recipe for soil, sand, silt and clay act for the flour, sugar and butter in chocolate cake (the bulk of the mix). A great cake, and a great soil, needs structure, which comes from organic matter. About 5 per cent of soil mass is organic matter, which stands in for the cocoa and egg. The last two ingredients are water and air. About 25 per cent of the volume of living soil is air. A cake without air is mudcake.

Planet Earth is really quite old, but our species, *Homo sapiens*, only branched off about 200,000–300,000 years ago. Compared to over 50 million human lifetimes at current life expectancy since the Earth was formed, that's only about 3750 lives.

We're a relatively new arrival on the planet, and we emerged in an ecosystem that had been hammering along for at least a good 3 billion years quite merrily without us. We are the result of all those species that came before us – including the microbes, which live in and on us.

Bacteria, archaea and fungi not only make soil, they also make us. Viruses played a part in our development, with vestiges of their DNA in every mammal. We are less than half ourselves: 57 per cent of the cells in our bodies are microbes.[9] It has been estimated there are 100 trillion individual bacterial cells in a single human body, which alters our genetic potential over a hundred-fold.[10] Our gut, in fact, is one of the most microbially dense ecosystems on Earth. These microbes don't just aid digestion. Gut microorganisms have been shown to play a role in our physical and mental health, including a diverse and unlikely array of human diseases such as obesity, psoriasis, autism and mood disorders.[11]

Everything is intertwined. It's easy to forget this when you're late for the 7.35 a.m. train, heading into the office to meet with a grumpy boss, with a pay packet that barely covers the rent, let alone a posh night on the town. It's all too easy to think, 'Soil? Phht!'

That's what we've done in spades for the last century. And, as we'll see in the next chapter, that's about to become a big problem.

To really thrive in the long term, humans have to realise that we are the result of long, historical accidents, which started with mass changes in the Earth's structure. We are the result of evolutionary chains that include millions of other species, both within us and around us. Each successive organism relies on the ones that have come before. We are the culmination of thousands of species of bacteria, billions of mutations, countless variations in genetic makeup, and aeons of geological time.

Getting a grip on how small and insignificant we are, and yet how powerful our actions have become, is paramount. Nature is bigger than us. Older than us. Better than us, despite our conceit that we can control it. When you mess with nature, you know you can't win. Not in the long run. The complexity can be overwhelming; the numbers can defy our comprehension. If we're humble enough, however, we can find our right place in the world. We can work with nature, not against it.

SOIL COMPOSITION TEST – IN A JAR

The ideal ratio of the three mineral components of soil is about 40 per cent sand, 40 per cent silt and 20 per cent clay.

Sandy soil can drain nutrients more quickly, as water passes through it faster. Clay soil can be too dense to allow much of the vital air that soil life needs, and can get more waterlogged – though all of that is also dependent on soil life.

To see how your soil stacks up in terms of its general sand, silt and clay composition is really easy. All you need is a jar, water and a ruler.

Find a large, tall, straight-sided jar that has a lid; ideally the jar should be about 1 litre (4 cup) capacity. Dig up soil from the patch you want to test, discarding the organic matter (sticks, dead grass and the like) right on the top. You want to test your soil right down to about 20 centimetres (8 inches) below the surface, so dig an evenly deep hole and mix up the soil in a bucket, taking multiple samples if you want to get a broad idea of your garden. As you mix the soil up, discard any rocks or large pieces of organic matter.

Fill your jar about one-third to half full with soil. Add enough water to come about nine-tenths up the side of the jar, leaving enough room to be able to shake it. Add half a teaspoon of dishwashing detergent if you have it, to help separate the soil components. Pop the lid on and shake the jar well for about three minutes.

Leave the jar on a bench, undisturbed, for a day or so. After that time, you'll find that the jar contents will have separated into layers. On the surface of the water will be organic matter. Below that, there should be water, probably discoloured with dissolved organic matter. The next layer down is clay, below that is silt – and below that is sand which, being heavier, should have sunk to the bottom.

Use a ruler to measure the total volume of the sediment, and the depth of each layer of minerals. For instance, if you have a total height of 10 cm (4 inches) of settled soil in the jar, you may find that the silt is 4.5 cm, the clay 1.5 cm, and the sand 4 cm. Convert each of these to a percentage by taking each layer, dividing it by the total height, then multiplying by 100. In my example, 1.5 cm of clay would be $1.5 \div 10 \times 100 = 15\%$. So clay would be 15% of my soil.

The U.S. Department of Agriculture has a fabulous soil pyramid to calculate the type of soil you have from these percentages – see http://tinyurl.com/soilcalc

The jar test is really handy for a quick, free analysis of what your general soil type is like. It's your own personal starting point. But remember, decayed organic matter and invisible life in soil can alter the way these minerals behave, because carbon and soil biology give soil structure, and can make minerals more available to plants. So, just because you don't have the ideal ratio, doesn't mean you can't grow great things. Regardless of ratios, soil life is fundamental, and should be the goal of all growers, no matter what the location.

CHAPTER 3

The Earth's Kidneys: When Good Soil Turns Bad

It's 2013, and I'm at a fish farm off the coast of southern China, where several species of grouper are sea-fattened for the pot. There are dead fish floating on the surface. The fish farm manager is barely interested, smoking constantly, flicking his cigarette butts into the water and occasionally glancing at his watch. The rectangular mesh enclosures apparently are full of live fish, not just corpses. Hundreds of fish farms stretch up and down as far as the eye can see. A worker dips a net into the water – through an oil slick, past some older decayed cigarette butts, and avoiding floating polystyrene. He pulls up a few live grouper for us to admire. Between the dead fish and the oil slick, I make a mental note not to eat seafood that comes from water like this.

In 2017, it was reported that about 81 per cent of China's coastline was heavily polluted – mostly with things like inorganic nitrogen, reactive phosphate, fluoride or oil.[1] That's a little better than when I visited the grouper farm in 2013. But the interesting thing about the pollution I saw was, simply, that I saw it. Most pollution, as with nitrogen and phosphate, you only see when you get a so-called red tide, when the

29

water turns russet with algal blooms from phytoplankton that thrive in the overly rich water. What happens in soil isn't always so apparent – but the waterways give us a clue.

Nitrogen, phosphates and fluoride aren't all that you can't see. One in ten of China's major river outlets is contaminated with DDT,[2] a terrible insecticide that is associated with liver and pancreatic cancer, a decrease in semen quality, effects on menstruation, and an increase in spontaneous abortions. It's classified as probably carcinogenic to humans, and there's evidence that pregnant women with high levels of DDT in their blood are more likely to have children who develop autism.

The thing is, these toxins in the water don't come from nowhere. While some are pumped straight out of poorly regulated factories, most comes through – or from – soil, or the interactions we have with soil, through activities such as growing crops and keeping livestock.

China's soil is a textbook example of what can go wrong in agriculture. Prior to the 1980s, China's agriculture was run predominantly on organic principles. That was mostly because it was done on a small scale, and there was no capitalist system to reward destruction of soil for immediate gain. Herbicide and pesticide use was moderate. Use of livestock manure (and humanure, composted human waste) was widespread.

Fast-forward only 17 years, and thanks to a massive increase in unregulated agriculture, China's official Report on the State of the Environment from 1997 describes the pollution of China's arable land (cropping land) as 'rather severe', with pollution affecting an estimated 10 million hectares (25 million acres) of soil.[3]

Another 17 years on, in April 2014, the Chinese government released another soil pollution survey of farmland, reporting that about 16 per cent of China's soil and over 19 per cent of farmland was contaminated.[4]

China's problems are manifold. They have less than 10 per cent of the world's arable land, to feed about 20 per cent of the world's population.[5] One out of every five hectares is poisoned by substances such as cadmium, nickel and arsenic, and a further 3.32 million hectares (about 8.2 million acres) of arable land is considered moderately polluted.[6]

The result of having contaminated soil in a country so populous means the food system is also contaminated. Official estimates are that China produces 12 million tonnes of grain contaminated with heavy metals a year.[7] So the land is poisoned, but they still use it – they probably *have* to use it – for agriculture.

Much of the contamination of soil is due to heavy metals. Much of it is from the output of cadmium, which really isn't a good thing to have in agricultural land. It causes bone and joint diseases. It can cause cancer. In the early 2000s, it was found that 10 per cent of all rice in China tested positive for cadmium.[8] While cadmium has been used for centuries in the porcelain industry, this can't explain the rise in recent times. Phosphates in fertilisers added to soil can also contain cadmium – and this is the actual culprit.

It's not only China that has poisoned its soil, of course. All countries have regions affected by mining or other disturbance. Lots of toxic chemicals that were previously stored safely underground have been discovered by humans, and brought to the surface – lead, mercury, uranium.

Cadmium has also posed a problem in other places, most notably the dairy farms of New Zealand, which imported cadmium with their phosphate fertiliser. Cadmium bio-accumulates in soil, building up in concentration over the years. New Zealand researchers found that super phosphate may also contain highly reactive fluorine, and isotopes of uranium, thorium and radium.[9]

Soil acts as the world's kidneys, filtering water and hanging on to lots of things we don't really want to ingest. The problem is, our soils are accumulating these things in our food-growing regions, and while the earth may hold back some toxins from washing down the creek, it can also release them to plants.

Cadmium is a problem, in part, because while it is toxic to plants, it's toxic at a lower level for plants than for humans. So vegetables, fruits and grasses don't show signs of sickness at the same concentrations that creatures that ingest them – humans and livestock – can.

The other problem is that it is cumulative, which means it just keeps adding up not only in soil, but also in our bodies every time we're exposed. This means non-toxic doses, added together, can have potentially serious health impacts. Imagine those heavy metals or radioactive elements in your milk, your meat, your cauliflower.

There are other agricultural poisons, too. Even copper, used as an antifungal agent in vineyards, can accumulate and affect vine growth. Brazilian researchers have found that copper that has been sprayed on vines can increase soil concentrations to such an extent that it poisons young vines when they're planted.[10] Nickel, while naturally present in soil in small amounts, can accumulate and become toxic not just to plants, but also humans. We've actually doubled the amount of nickel that naturally cycles through soils simply by burning fossil fuels.

Europe has its own problems – not as marked as China, and more varied than New Zealand. According to a 2016 report examining the concentration of heavy metals across 22,000 sites in Europe, including arsenic, cadmium, chromium, copper, mercury, lead, zinc, antimony, cobalt and nickel:

> An estimated 6.24 per cent or 137,000 square kilometres [53,000 square miles] of agricultural land needs local assessment and eventual remediation action.[11]

That means that one in every sixteen hectares in Europe's growing land, their usable soil for food, needs healing.[12]

Worldwide, heavy metals alone cost about US$10 billion a year in economic terms for loss of production and remediation.[13] It's a big problem, of course, if it's your land, or the land your food is grown on.

The United States also has issues with contaminated soil and its implications downriver, but as in Australia, because there's less industry and more open space compared to much of Europe and Asia, most of the land affected is being harmed by more broad agricultural products, rather than heavy metals. According to the US government:

> Agriculture is the leading source of impairments in the nation's rivers and lakes. About a half million tons of pesticides, 12 million tons of nitrogen, and 4 million tons of phosphorus fertilizer are applied annually to crops in the continental United States.[14]

Roughly half of that fertiliser ends up in waterways.

There has also been the widespread use of arsenate of lead in many fruit-producing regions until the 1970s or so. Added to that is the legacy of lead in fuel, which has accumulated in areas close to heavy traffic, until its use in fuel was phased out and banned in most countries by the early 2000s.

DDT is also found in some Australian and American farmland, something for which we can thank our grandparents, when they used it on agricultural sites and domestic plots alike. DDT has a half-life in soil of between 22 days and 30 years;[15] half-life is the time it takes for half of an amount to degrade. It's considered persistent in the environment, meaning it takes ages – like decades – to degrade to safe levels.

Of course, our farming contaminants aren't just a worry on the land, as we saw in China. While topsoil does act as a filter, trapping a lot of heavy metals and contaminants, many of these contaminants do end up downstream.

The U.S. Geological Survey, an arm of the federal government, declares:

> Pesticides are widespread in surface water and groundwater across the United States. For example, at least one pesticide was found in about 94 per cent of water samples and in more than 90 per cent of fish samples taken from streams across the Nation.[16]

In New Zealand, two-thirds of all rivers are now unswimmable, thanks to nitrogen run-off, mostly from dairy farms. And three-quarters of their native freshwater fish species are under threat of extinction. Agriculture, what we do to soil to grow food, is almost entirely to blame.

So, some of our soil is laced with DDT. We've over-fertilised and caused the deaths and sickness of river systems. Between cadmium, mercury, nickel, arsenic, cobalt and lead, we've also pretty much thrown all the dangerous heavy metals we could find at soil, along with a few radioisotopes. Some of these contaminants are way more toxic than others. It only takes seven tablespoons of lead, for instance, to poison a hectare of soil.[17]

Not all poisoning of land is poisoning us quite so directly, though. Much of the contamination of cropland comes from salt. This increased salinity is caused by rising water tables, usually as a result of removing trees from a landscape. As you take away trees, and their deep roots die with them, the subterranean water rises, carrying salt with it. Most land-based plants have a low salt tolerance, so they soon die off, or at least get quite sickly. If you fly over Western Australia and look down at the wheat belt, you'll see massive white blooms on the earth 10,000 metres below. It's not the state's famed wildflowers, though, that you spy from that far up. It's saltpans that dot the landscape – essentially buggering up 1 million hectares (2½ million acres) of agricultural soil.

One million hectares of soil, lost from being productive land, no longer able to grow food for humans to eat. Lost to salt.[18]

Between the poisoning of land, and erosion, it's estimated that the world has lost a third of its cropland in the last 40 years.[19] We're fast depleting the very thing that feeds us and filters our water. We've used the land as our dumping ground, without fully understanding what we're leaving behind in our wake.

What we're losing isn't just sand and silt and clay. What we imperil is an impossibly complex ecosystem that we're only on the cusp of fully understanding, but upon which we all rely.

CHAPTER 4

Plants Don't Eat Dirt:
The Underground Economy

Letitia Ware grimaces as she tries to insert a spike into our paddock. This simple instrument, a long, double-handled metal spear with a dial at the top, will give us a measure of our land. The higher the needle goes on the dial, the more compacted the ground is. It doesn't look good when the needle on the meter reads off the scale. Compacted soil is not very good soil – but that's not the end of our soil's troubles.

It is 2011, and our first year owning Fat Pig Farm, our 70 acre (28 hectare) plot that we want to convert from grazing land to a mixed holding, raising pigs, cattle, sheep and a market garden. The gentle slope of our future garden sits at the valley floor and faces north, to capture the all-important sun.

When we bought the place, the grass was emerald green and almost waist high. To our untrained eyes, it looked like really good growing land. Better than the scrubby hill that rises behind our farm, nicknamed Bandicoot Hill, which takes its moniker from the local marsupial, because the soil's so poor you can't even grow bandicoots on it. Our place is an end point of Bandicoot Hill's topsoil, as it washes downhill over the years.

Despite our better soil, Letitia's face still turns ripe plum purple in the summer sun, as she tries to drive this spike into the ground.

We take a sample of our farm back to her lab, and under the microscope she shows me what is there. Mostly, there are rock minerals – the silt that dominates our corner of the region, and some clay. And um … not much else. Eventually she finds some bacteria. And finally a nematode, which I mistook for a tiny worm. Under the microscope, soil from the best spot on the farm was pretty much lifeless and dead. Yes, I see a nematode. I see a few bacteria. But, really, I don't see much.

But when we looked at Letitia's soil at her garlic farm, all we could see was life. It teemed with thousands of oval bacteria, hundreds of nematodes, and was littered with fine filaments of fungi. For the first time, I witnessed the living nature of soil.

It must have been wondrous for early scientists to see things not visible to the human eye. It must have blown their minds. I grew up with a microscope and a chemist father, yet still, seeing those living things in the earth – even the few in our soil – filled me with wonder.

It still fills me wonder. That there can be 10 billion living things in a teaspoon of healthy soil[1] is almost beyond comprehension.

It seems too inconceivable to be true that, in the amount of soil you can scoop up in a single go with your bare hands, there can be:

5000 insects, arachnids, worms and molluscs (from up to 500 species)

100,000 protozoa (from up to 500 species)

10,000 nematodes (from up to 100 species)

500 metres of plant roots (from up to 50 species)

100 billion bacteria (from up to 10,000 species)

100 kilometres of fungal filaments (from up to 1000 species)

plus algae, archaea and more[2]

The more scientists have looked, the more they've realised that soil, like us, is driven by microbes. They've realised that the world is governed by things too small to see with the naked eye, and too numerous to count, except by using projections and algorithms.

We've known about soil microbes since the invention of the first powerful microscope, in the late 1600s. And we've known that the best soil is usually dark, has a certain texture, is aided by decomposing leaf litter and more. But it's only recently that soil science has started to catch up with ancient wisdom, validating the use of compost, the worth of comfrey tea, the majesty of worms. It's really only in the last two decades, thanks to a return to thinking of soil as a living organism (as well as some super-duper advances in scientific techniques), that we've truly started to understand the role of soil.

Once worshipped as the nurturer – Mother Earth – soil began to be thought of in base terms. In the early 1800s, people worked out what you needed to grow plants bigger and faster. Pioneering work from scientists such as German chemist Justus von Liebig demystified chemical pathways in soil, and discovered the pivotal role of major plant nutrients nitrogen (chemical symbol N), phosphorus (P) and potassium (K), and how each was essential for plant growth. Prior to this, humus, the dark organic matter in good soil, was considered to have mystical properties, a kind of god-like life force; something unknown and unknowable.

Basic chemistry stripped soil of its status as a deity, and endowed humankind with the ability to manipulate crops. We'll look later at what has ensued over the last 100 years in particular, but suffice to say, what a keen gardener knows now, they also knew in the late 1800s – that compost is king, complexity seems to create resilience, and that a well-grown vegetable tastes of the soil in which it was grown.

But it's probably most important to remember that soil isn't soil unless it comes in contact with plant roots. Soil with a scant few microbes, separated from plant roots by distance or time, is dirt. And it's here that all that subterranean life comes in.

I used to think, before I met Letitia, that plants ate dirt. I used to think that, somehow, a plant's roots went into the ground and dug around looking for things to eat, like some kind of mystical creature that can consume rocks. I pictured their roots like straws, drawing up what they wanted from the earth. Sure, I knew there'd be a chemical component to this, a way for a plant to extract calcium, say, from sand – but I never thought of the conduit as being a whole different form of life. Or, more accurately, billions of lives. These lives might constitute only 0.5–1 per cent of soil mass,[3] but they pack a punch collectively.

According to the European Commission, a hectare of healthy pasture has five tonnes of microbes living in the soil, equivalent in mass to about 100 sheep. That same hectare of grass would only be able to feed about 20 sheep.[4] So, soil supports five times more life below ground than above. Bacteria dominate the landscape, even if we don't see them. There's nowhere humans have been on Earth that doesn't show some sign of bacteria, now or in the distant past. And most bacteria live in soil. The 'germs' we were taught to fear are discovered to be working night and day to gift us life.

Nadia Danti is gesturing wildly with her hands. Her Italian heritage is obvious, and it is one she shares with a juvenile broad bean she's just plucked from the earth. The fragile, bright-green seedling is getting waved around like a magic wand as Nadia talks about the garden she's in charge of, where Letitia Ware once nearly busted her gall checking the compaction.

It's 2017, and as the market gardener on our little farm, Nadia is hosting part of the farm tour. When people come to see us, and eat in our little dining room, they always get a look at the garden. They also get told things a lot of them probably didn't know before.

I'm still learning stuff, too. As Nadia talks about her role in the garden, as she explains that she is responsible for growing soil, I can feel a ripple of disbelief amongst the crowd. She's talking about that massive, functional soil ecosystem. Look after the soil, Nadia tells us all, and the plants will look after themselves.

Which isn't quite the full story. The plants still need weeding. They still need water. But in essence, it's true.

Holding aloft the delicate broad bean, she points to nodules on the roots. Tiny white and pink baubles are tucked up next to the fine hairs on the dirt-clumped roots. These nodules are where this family of plants – legumes, in garden speak – fixes nitrogen. Certain plants, it's been known for centuries, can help feed the soil with something that really kicks in vegetative growth. Legumes such as peas and beans capture nitrogen from the air and hold it in these nodules, which help replenish soil's fertility. They make the ground better for themselves, and the plants that follow.

I'd heard of this before. Except it's not the plant that does the storing of nitrogen. As Nadia talks, she shows the colour difference between nodules, and explains that a plant is just a home. The thing that is taking nitrogen out of the air, feeding it to the broad beans, and fixing it in the soil for future generations, is bacteria. Rhizobium bacteria have found a site, and a willing host, that allows them to trap nitrogen and store it. Billions of them have turned some of the nodules pink, a sign that they've snaffled this essential plant nutrient from the atmosphere, and have stashed it, like the best wine or cheese, under the ground.

The red colour comes from leghaemoglobin, a leguminous form of our blood protein, haemoglobin. This plant-provided protein rises in colour in proportion to the amount of nitrogen stored in the bacteria. And just like human haemoglobin, its role is to regulate the oxygen supply to bacterial and root cells.

Why do the rhizobium bacteria do this for the plant? The plant gets nitrogen, but what do the bacteria get? Well, they do it because the sugars a plant makes through photosynthesis can be exuded – essentially dribbled out – into the soil. There can be thousands of carbon compounds that a plant oozes from its roots, where microbes cluster like fish at a jetty. They gather together, working sometimes in concert, sometimes in competition, to find the things the plant wants and needs, and they get fed into the bargain.

Bacteria bunch minerals together in little orbs, called micro-aggregates. These tiny bundles of soil particles are held together with polymers – like sticky glue – that the bacteria create. These micro-aggregates help give soil structure. All the energy for the bustling work – and the carbon to build the polymers – comes from the plants.

The entire plant/soil system is based on an underground economy. Forget the idea of a root sucking up food from the ground – the real vacuuming is happening above ground. Green living plants are constantly hoovering the air, drawing in carbon dioxide, transforming it into sugars and piping them downward. While the plant is obviously using that energy to aid its own growth, fight diseases and ward off pests, plants can also trade at least one-fifth of the sugars they create with the microbes around their roots; some send up to 80 per cent of the energy they photosynthesise underground.[5]

Only a few plants have large numbers of rhizobium bacteria around their roots. So what does a non-leguminous plant get from microbes, in exchange for sugars? After all, the plant already makes its own food, so it doesn't need any carbohydrates. Well, remember van Helmont's potted willow, which grew 75 kilograms (170 pounds) in five years, while only using 57 grams (2 ounces) of the soil? Those 57 grams are the micronutrients the plant needs to grow and thrive – beyond carbon, oxygen, nitrogen and hydrogen, which all come from the air. A plant relies on microbes to do some of the finding of those nutrients, and to chemically transform these into a digestible form.

A plant can't move through soil, except by slowly extending its roots. But other things can move through soil much faster, and transport nutrients much more efficiently. They can also travel much further. If all the roots of wheat in a 1 hectare (2.5 acre) field were laid end to end, they would stretch more than 46,000 kilometres (28,500 miles) in length[6] – which is greater than the circumference of the Earth. But fine fungal threads, called hyphae, that have a relationship with the plant, can stretch over 2000 times as far.[7] It's not that they stretch out in a single line for kilometres away from the plant, but rather that they crisscross the soil in three dimensions, finding the elements that the plants want.

The area around a plant's roots – the bit where Nadia's nitrogen-fixing bacteria were clustered – is called the rhizosphere. It is the root/soil interface, where most of the nutrients are sourced, and most of the living microbes cluster. But further away – in fact, way further than the bacteria can move, or the plant roots go – are other microbes. And these are also important, thanks to the presence of fungi.

In many ways, fungi are the heavy lifters in soil. About 90 per cent of plants, and pretty much all trees, have relationships with fungi.[8] We tend to think of fungi as decomposers, because we see mushrooms – the fruiting body of certain fungi – on dead trees. Many work in concert with plants, forming tight relationships with not only one plant, but perhaps many. These underground fungi form networks, like microscopic telephone wires, or even superhighways. These fine tendrils of hyphae are often too small to see, which is perhaps why they have long been overlooked.

We don't really understand fungi yet, despite sharing up to 80 per cent of our DNA with them.[9] The mushrooms you eat or see in the forest are just a more visible, tangible expression of fungi. Moulds and fungi can be found in the most extreme environments. They are thought to be the most diverse biota in Antarctica, even under its permanently frozen lakes;[10] they also exist in really acidic environments (they actually help make vinegar). Fungi have even been found growing around Chernobyl after the nuclear energy plant's accident, despite the ionising radiation. Fungi live off other things, be those other things alive or dead; hence it is likely they evolved after bacteria and other life appeared.

So what are fungi, if they are not simply mushrooms? Yeast is a fungus. The white fuzz on the outside of your Brie is a fungus. Pretty much everywhere there is life, there's a fungus ready to take advantage of such life, and break it down once it's dead. On Earth, there are estimated to be 3.8 million species of fungi,[11] doing everything from decomposing trees to causing athlete's foot. They're rotting the inside of organ pipes in ancient churches, fermenting your beer, and raising your sourdough. Only a tiny fraction – 120,000 species or so – have been described by scientists.[12] Some live off dead leaf matter. Some live in our intestines; a study in 20 people from varied backgrounds found fungi from 85 different genera, or groupings, living in their gastrointestinal tracts.[13] Some inhabit the guts of bees – with the nature and type of fungus varying according to social status within the hive.[14]

The vast majority of fungi spend at least part of their life cycle in soil.[15] Despite this, it's only very recently that we've started to fathom their role in soil. One of the most important is that fungi provide much of the 'glue' that holds soil together, a thing called glomalin. And that wasn't even discovered until 1996, by an American soil scientist, Sara F. Wright.

Glomalin is a glycoprotein in structure, we think; we still don't know for sure. It's easiest to think of glomalin as having binding ability, similar perhaps to the eggs in our great chocolate cake. It's part of the organic matter of soil. For a long time, we didn't know glomalin was even there in soil, simply because the way you pull apart soil to get to it is pretty severe. 'It requires an unusual effort to dislodge glomalin for study: a bath in citrate combined with heating at 250°F [121°C] for at least an hour ... No other soil glue found to date required anything as drastic as this,' explains Wright.[16] To extract glomalin, you have to buffer the soil, and then cook it under extreme pressure (pressure-cooker style). The process alone is so brutal that it can denature most of the other things scientists like to study.

Glomalin holds between one-third and one-quarter of the world's soil carbon, yet it still seems strangely absent from so many discussions on soil or fungi. (Merlin Sheldrake's fabulous fungal celebratory tome, *Entangled Life*, for instance, has no reference to it.[17])

The way glomalin is made is still also a mystery. But we know the basics. Plant-associated fungal hyphae are about 1/60th of the width of root hairs, and there can be 10 kilometres (6 miles) of them in a teaspoon of soil.[18] Yep, that's right, they're tiny. But their thin size makes them even more remarkable, because nutrients are passing each other in both directions. It's estimated that most hyphae only live for days or weeks, before being sloughed off into the soil. In that time, the hyphae are transporting nutrients (often phosphorus), dissolving minerals and feeding plants. The hyphae would make great food for many other subterranean inhabitants – so they protect themselves with a resilient shield made out of, you guessed it, glomalin.

Glomalin's way of gluing soil together means that sand, silt and clay can clump up in a way that still allows the passage of air in and out. It allows soil to hold and to drain water better, and it's particularly important in the rhizosphere. Bacteria create those soil micro-aggregates we met earlier, while glomalin creates macro-aggregates. It's like lining up a bunch of feedbags around the roots so the plants can access what they want, when they need it – but only if the microbes are healthy.

Glomalin is far, far more resilient in soil than most other carbon elements, with estimates suggesting glomalin can remain for up to 50 years in undisturbed soil.[19] Glomalin is what gives soil the tilthy texture we love, and helps stop soil being blown or washed away. This little-understood glycoprotein is the long-term storage method that helps soil, through its association with plants, look after itself.

Consider this. For 12,000 years we've been altering soil drastically as we grow food. It's been over 5500 years since humans invented the plough[20] to help turn over the soil we farmed. For over 300 years we've known that bacteria live in water and soil, and in us. A hundred years ago we industrialised the burning of natural gas to fix nitrogen to make fertiliser, which in turn has produced about half the calories in food you eat today. Fifty years ago we put a man on the Moon and collected samples of the dust he found there. But it's only 25 years since we discovered glomalin.

We really haven't been paying proper attention to soil.

The hyphae from fungi can weave through fissures in rocks, actually cracking the rocks further, using acids to break them down into fine minerals. You don't have to wait for a glacier to roll over the land to release nutrients from rock. Fungi, and to a lesser extent other soil life, can do it for you. Which is rather good news for most home gardeners and farmers alike.

No-one is quite sure how hyphae can have water and minerals going one way, from the terminal fungal ends to the plant roots, and at the same time have sugars passing the other way, to fuel the fungi's growth. It's like a two-way street, with no dividing line. Nearly 200 years after von Liebig simplified soil science and disputed any vitalist life force in humus, we still don't know how fungi work in soil.

Research on soil fungi is increasing, and every revelation takes us further into the fantastical. Soil fungi can be used to clean up heavy metals.[21] They can draw the proteins from salmon carcasses down into the ground, and move them within reach of trees that then utilise the salmon's nitrogen in their own structure, high up in the branches. Soil fungi pass on messages to other plants to warn them of predators or disease, which allows other plants to put out pheromones, or change their chemical structure, to ward off attack.[22] They're like underground conduits, allowing the passage of food, messages and moisture.

In fact, work by Suzanne Simard and others, particularly in the United States, has shown that trees form these incredible networks, aiding not just their own species, but also other species.[23] They share water and sugars, depending on the season. And when a tree is likely to die, it will transfer much of its available carbon to younger trees, even those of a different species, using fungi in concert with other microbes. Of course, like all things that arrived later on the evolutionary scale, fungi are mediated by things that arrived earlier. Things like soil bacteria and archaea.

While 90 per cent of soil's microbial life in a healthy forest is fungi, in our gardens, paddocks and cropping land it's usually about 10 per cent.[24] Trees really need fungi, because to make lignin – the stiff bit of a tree that allows it to stand up tall – they need lots of the elements that fungi can provide. Meanwhile, the fungi need trees (or other photosynthetic plants) for food. In a true cyclical fashion, the only way to break down dead trees is through the action of white-rot fungi. Fungi, alone amongst the microbes, can return a tree to useful components ready to be cycled back into soil, and eventually new trees or plants.

Many growers now recognise that all healthy ecosystems need subterranean fungi, probably in higher proportions, and much higher quantities than we have in our gardens and farmlands. Some suggest that a ratio of 50 per cent fungi to other microbes would be ideal – way more than the estimated current 10 per cent.[25]

Fungi aren't just eating and passing on sugars and messages. And they're not just finding minerals lying around in the soil, ready to pass back to a plant. They're mining them. A fungus can seek out rocks in the soil, secrete acids onto them, and then drill, effectively, holes in them. Through these holes in the rock, the hyphae then extend and extract the minerals a plant needs.

Along with being miners, fungi are also murderers. They can predate on little collembolans (sometimes called springtails because of their ability to leap great heights), colonise the collembolans' bodies, and mine them for their nitrogen.

Fungi are also cowboys. They can lasso nematodes. They create a loop in their hyphae, attract nematodes with chemical signals and then, when the nematode is in the loop, they increase the size of the cells in their hyphae and essentially constrict the nematode, before releasing enzymes to digest the nematode. They also make nets to trap them, like underground fishers. The fungi then give most of the nitrogen they harvest to plants to use.

Of course, the springtails can also eat fungi, too, depending on the species. It's really dog-eat-dog down there. And all the while, as they cycle nutrients, making them more available for plants, springtails actually play a vital role in spreading fungal spores, too.

Microbes can work in concert with one another, with bacterial and fungal networks actively seeking nutrients on behalf of plants, building resilience, storing carbon, cycling minerals. They can help plants to share nutrients, even young growing plants of different species. Sometimes, the networks can even gang up.

One example of this is a tree we planted in our heritage apple orchard nearly a decade ago. This sapling is about head high on me, about six feet, or just under two metres tall. It's very small for its age. The tree has never fruited. It appears stunted. At its base, it's only about three centimetres wide, a bit over an inch. Compared to the trees around it, which are 80 years old, gnarled and wider at the base than an axeman's thigh, this one looks puny.

We've made a mistake planting this tree here. It's suffering from something called apple replant disease, a malaise that occurs with all pomme fruit – apples, pears and quince – when an old tree is replaced with a new one.

It turns out that the microbial community that forms under an apple tree is very committed to helping it thrive. The fungal networks seek out calcium, the bacteria fix nitrogen, and nematodes crawl through the soil gobbling up the bacteria. Soil mites act as ecosystem engineers, keeping the populations in check. As the apple tree's roots penetrate the earth, this community grows with it, nurturing it, feeding it. They form a multi-species super-organism with one goal: to nourish the apple tree. But once the original tree is cut down, something odd happens. The ecosystem under the soil turns deadly.

It just so happens that the subterranean community gets a bit too attached to the original plant. Nematodes, which hatch as a result of exudates from the former apple tree's roots, live in the soil and attack the new plant's roots. Pythium, an oomycete (fungi-like multicellular organism), strips root hairs from the new tree. The whole process is little understood, but evidence points to a combination of fungi, oomycetes, nematodes and bacteria working in concert to starve the new tree of nutrients. Apple replant disease can exist in the soil for up to 15 years after the old tree dies. For a new apple tree, there's something dangerous in the deep. Our subterranean soil life fell in love with the first tree it nurtured and doesn't want another tree of the same species to grow where the last one was.

Why does this happen? Is it a way of making sure that apples that fall from the original tree don't thrive, thwarting their attempts to take over from the old parent tree? Is it a way of making sure that the only really viable seeds are ones that have been taken further from the source? Nobody really knows.

What we do know for sure is that the disease is microbially mediated. If you sterilise the soil, fumigate it using chloropicrin or other similarly biologically disruptive chemicals, you can reduce the chance of getting apple replant disease to virtually zero (though you'd also be wrecking the ozone layer as you did so). This suggests that it's the life of the soil, not localised depletion of nutrients, that causes the disease. Good results come if you dig up large volumes of soil from around the site of the new tree, and bring in fresh soil from elsewhere (other than an orchard!).

The ecosystem under the ground drives what happens above ground. It can happen in ways we don't fully understand, and in ways, as with apple replant disease, that seem to defy logic.

It's far easier to imagine an ecosystem above the ground than below, with plants, small mammals, lizards, birds and insects all coexisting in the geography and watercourses that support it. There are herbivores and omnivores and carnivores, a million connections.

Below ground, the ecosystem is way more complex, thanks to more species, and a three-dimensional living space. But it, too, has herbivores, omnivores and carnivores, as well as a host of other really cool and complicated interactions. The problem is, it's unseen – and until relatively recently, unseeable and unknowable.

While leaps have been made in our understanding of microbes, of plant growth, of nutrition and all the complicated chemical inter-actions, there's still a lot that we don't yet know.

Partly that's because what happens below ground is very hard to measure. In 2018, in the magazine *Rhizosphere*, researchers questioned whether we can actually measure exudates – the chemicals that are leached out, or pumped out, at the ends of plant roots. Being underground, this kind of science is almost impossibly hard to conduct. The name of their research paper says it all: 'Sampling root exudates – Mission impossible?'[26]

Sadly, it's still an area of science we don't put a lot of effort into, and with every passing year, the fantastical seems to replace the simple. At first it seemed amazing that roots could penetrate through soil with a force of 100 psi at the tiny tip of the plant. Now that is taken as a given. Now we know that the root tips of plants also exude organic acids and phytochemicals such as flavonoids, carotenoids and polyphenols, as well as sugars and amino acids (the building blocks of protein).

So many root exudates are being discovered that nobody seems to want to guess at how many we'll find. Plants can make at least 100,000 of them.[27] Some are used to suppress disease or fight pests. Some work as attractants, bringing in soil mites, and others act like hormones, triggering growth in microbial life. Some are dribbled out to get the microbes to actively attack a neighbouring plant in times of drought. Some are simply, we think, food for the underground ecosystem.

Those exudates help plants in their root-to-root exchange, as they communicate about swapping nutrients, or to warn of pests. They are also vital in root-to-microbe exchange. The exudates change in response to toxicity – plants actually put out detoxifying exudates in the presence of high aluminium levels to reduce the harmful effects of the metal. Plants also change their exudates in low-phosphorus soil to try to make more phosphorus available, altering root growth and architecture, as well as microbial populations at the same time.

Plants are influencing, directing, managing things. They actually farm microbes in, on, and underneath themselves.

Plants can only do it, though, in healthy soil, with the enormity of underground life. Soil is the medium, life is the result.

And while plants make thousands of chemicals for the soil, the soil also makes thousands of chemicals for the plants. The reason this matters is because what a plant is exuding and receiving is also what it contains. These thousands of chemical compounds, which are so far proving unmeasurable, are the same ones in their structure – nutrients that help the soil and the plants to thrive, and which can do us good, too.

Ultimately, I was wrong: plants don't eat dirt. What I first heard of from Letitia, and then from countless other sources, is that up to 90 per cent of what a plant needs from soil comes through the action of microbes. Microbes make it happen. They communicate, somehow, with the plant about what the plant wants; the microbes then source it, transport it, and make it available in a form the plant can use.

Despite comprising less than 1 per cent of soil mass, it is soil microbes that actually matter when it comes to what a plant can, and will, get to eat, after that initial sugar rush from photosynthesis.

Plants and soil can only coexist. Like us, plants came into a world dominated by microbes, and evolved in their presence. Multiple complex interactions can be hard to understand. Yes, we can grow plants in places where the soil is negligible. Or even in water that we add nutrients to, as in hydroponics. But we need to have the humility to recognise that this is not how nature makes the best food for us to eat, and that just because we can't quantify all the different microbial interactions between plant and soil, doesn't mean they're not happening, or that they're not important. The result of ignoring soil life is inevitably to produce more hollow plants. We have created plants that are shadows of their possible selves, in terms of the complexity of their makeup, the resilience in their being, and the vitality of elements and molecules that they could, and probably should, be hosting.

We often hear that we are what we eat. In terms of plants, we are what a plant eats, too. And if you eat meat, you are also what your animal eats. Essentially, you are what you eat eats. And all of this, it must be said, comes down to the quality of soil, and the abundance of tiny microbes in it.

A plant can contain more minerals and useful molecules – micro-nutrients such as polyphenols – if it can readily obtain the building blocks from soil. If only there were an easy way to tell if healthy soil is making a difference to the food you eat, beyond the lab. Later, I'll show you that, yes, there is. And it doesn't cost a cent.

But first, let's take a look at some of the other unexpected things that soil life does for us, beyond feeding us and helping us breathe.

SOIL YOUR UNDIES

Those billions of living things that make soil active are mostly invisible to the naked eye. But you can test your soil in a general sense to see how active and vibrant your underground ecosystem is. This simple test, developed by Bill Robertson, a soil scientist at the University of Arkansas, is called Soil Your Undies because, well, you get your underwear dirty.

You need two identical pairs of washed cotton briefs. Bury each one in a different part of your garden. Perhaps one in the garden beds where you grow your vegetables, and the other in a spot with differing biology, such as a corner of the lawn, or under a tree. Dig a hole 5 cm (2 inches) deep, and bury the underpants so they lay flat. Cover them with the soil you've dug out (leaving just the waistband exposed so you can find them again later) and wait six weeks.

At the end of six weeks (hopefully you marked where you put them ...) dig them up and compare the pair. Biologically active soil will have eaten through those cotton jocks fairly effectively; the fabric (essentially cellulose – microbe food) will be barely held together. More lifeless soil will mean your undies are still in pretty good nick – though perhaps too far gone to re-use as a garment.

CHAPTER 5

Here, There and Everywhere: The 'Old Friends' Hypothesis

Mary O'Brien had no idea what she was about to discover when she injected an otherwise harmless soil bug, *Mycobacterium vaccae*, into seriously ill cancer patients at Britain's Royal Marsden Hospital in 2004.[1] An oncologist (cancer specialist), O'Brien thought the bacterium may help the human body in its fight against a form of lung cancer. She didn't just do this on a whim, of course. Positive-sounding results with the bacterium on tuberculosis, as well as the suggestion of heightened immune responses, had hinted it could be helpful.

Patients were given either chemotherapy alone, or chemotherapy with the bacterium. And *M. vaccae* proved useless for curing cancer. It did nothing at all to help clear the patient's tumours, or extend their lives. So the trial could be considered a failure.

Except it wasn't.

M. vaccae didn't work in the way O'Brien expected; it didn't prolong life. But it vastly improved the quality of life for the patients who were given the bacterial treatment, versus chemotherapy alone. The soil microbe didn't cure cancer as hoped, but it cheered the patients up. It helped them think more clearly, and they felt happier and more positive. They also felt less pain.

It turns out that the common soil bacterium *M. vaccae* acts like an antidepressant, boosting the levels of happy hormones, serotonin and norepinephrine, in humans and other mammals. It also reduces stress. In one study, published in the *Proceedings of the National Academy of Sciences* in 2016, injections of *M. vaccae* prior to a stressful event prevented a stress syndrome similar to post-traumatic stress disorder (PTSD) in laboratory animals.[2]

Christopher A. Lowry, an integrated physiologist from the University of Colorado, who co-authored the 2016 study, demonstrated that *M. vaccae* can have all kinds of effects. 'It seems that these bacteria we co-evolved with have a trick up their sleeve,' Lowry says.[3] Apparently they put out a lipid, a fat, which has a positive effect on the immune system, and also shows an anti-inflammatory response in the brain. Lowry is interested in whether he can create a 'stress vaccine' from the bacteria. An aid for human PTSD is one major goal.

Research suggests that ingesting the bacterium is just as good as being injected with it[4] – and because *M. vaccae* is everywhere in most healthy soil, every time you eat a piece of lettuce from your garden, you're probably getting the benefit of it. Breathing in the bacterium has proven to be a great mood lifter, too.[5] No wonder so many people relish the uplifting effect of working in the garden. And eating its fresh produce.

Like many microorganisms, *M. vaccae* fits the 'Old Friends' hypothesis. That is, because we evolved with bacteria, instead of treating them all like they'll kill us, we are actually *designed* to have them in us, and on us. Bacteria aren't all bad; it's just that some of them are bad when they're in the wrong place. For instance, having a large number of bacteria from your bowels getting into your mouth can make you rather sick. And yet, not having faecal bacteria such as *E. coli* in our colon is a recipe for disaster, because they're vital for helping us break down and digest food. They're not meant to be in the start of the gastrointestinal tract, but rather the middle and the end. Also, not having the entire range and thriving populations of bacteria that we are designed to have can lead to autoimmune responses, such as many of the allergic reactions we're seeing in heightened numbers today.

So, soil can improve our mood, but what else can it do for us? Well, soil – or at least what's in it – can also improve our ability to fight disease. Take the antioxidant L-ergothioneine, which is only made by soil fungi and soil bacteria. L-ergothioneine is widely regarded as having a role in slowing the human ageing process and reducing the inflammatory response – a great thing for heart health and overall health. One of the best ways to get it is through eating red meat (particularly liver) or mushrooms, but it also exists in some crops, such as black beans and oats. An experiment at the Rodale Institute in the United States showed that oats grown in fields that weren't tilled, so the fungal networks weren't broken, had 25 per cent more L-ergothioneine than oats grown in tilled fields.[6] Considering how tilling, ploughing, can break up fungal networks, this probably isn't a surprise.

'For a long time, [scientists] have been obsessed with the idea that there are things in the soil that are trying to kill us,' Rob Knight, a microbiologist at the University of California San Diego, told *The Washington Post* in 2019,[7] pointing to tetanus as an example. But we now know this is actually looking at things the wrong way.

There's abundant evidence from developed nations showing that urban children have more autoimmune diseases than those who are exposed to soil bacteria. Termed the 'hygiene hypothesis' in 1989,[8] the theory has been that many of our allergic reactions, and the diseases associated with them, are the result of being too far removed from microbes, particularly those from soil. The idea is that because we've evolved with exposure to countless bacteria, archaea and fungi, when these are removed from our daily lives, our bodies suffer.

We know that if you are exposed to a more diverse array of bacteria, from soil and from animals, starting in the first weeks of life, then you are less likely to have faulty immune responses – conditions like asthma and allergic rhinitis (allergic reactions in the nasal passages).

In fact, recent research suggests that the soil and the gut microbiome are intertwined. In a 2019 paper, 'Does Soil Contribute to the Human Gut Microbiome?', lead researcher Winfried Blum unequivocally states that 'the soil contributes to the human gut microbiome –

it was essential in the evolution of the human gut microbiome and it is a major inoculant and provider of beneficial gut microorganisms'.[9] In other words, soil gave us our bacteria – the complex internal ecosystem that allows us to thrive and survive. By removing ourselves from soil, and removing plants from microbially active soil, we limit our own ability to fight disease.

Blum compares the complex interplay of organisms in the rhizosphere around the root of plants to what happens in our guts. Some things feed us; some are fed by us. Some pass messages – and some can, if present in the wrong place or wrong amounts (or both), make us ill.

The hygiene hypothesis has been used to explain everything from peanut allergies to eczema, from hay fever to ulcerative colitis. Population studies have shown that the more soil microbes you've been exposed to, the more likely you are to have a stronger immune system.[10] Playing with dirt has never been more scientifically accepted.

These days, the hygiene hypothesis – which some were arguing meant you shouldn't wash your hands – has been usurped by Old Friends. Playing in the mud, being exposed to lots of non-harmful bacteria, of the kind we are evolved to encounter in soil, is like hanging out with old friends. It's not dangerous bacteria we need to be exposed to, but rather a wide range of harmless and potentially positive bacteria – the kind we mostly find in soil.

So, soil microbes can make you happy. They can reduce stress. They can reduce inflammation, and ameliorate some of the effects of ageing. It's likely that exposure to healthy soil, in the way our forebears were, can help reduce autoimmune diseases.

Soil microbes can also save your life. This is possible because the whole underground ecosystem is a bloodbath. Bacteria are being eaten by worms, fungi are attacking mites, mites are eating protozoa. But some of the most powerful agents for attacking bacteria are – wait for it – bacteria. In order to survive in an environment in which everything wants to eat you, some bacteria have come up with very good strategies to avoid being eaten by other bacteria. Hence the rise of antibiotics.

Actinomycetes, a general group of bacteria that are capable of breaking down more complex compounds, such as chitin from insect shells, form the basis of antibiotics. In fact, over half of the antibiotics used in human medicine – including aureomycin, chloromycetin, kanamycin, neomycin, streptomycin and terramycin – come from soil.[11]

Actinomycetes are also one reason soil can smell so good. That earthy, musky scent of freshly exposed soil comes from metabolic end products called geosmins that originate in actinomycetes and other soil microbes. If you've ever picked up a handful of healthy soil and smelled the glorious aroma, almost been tempted to taste it, that's geosmin. Geosmins give the earthy taste to some vegetables, too, such as beetroot – and the human nose can pick them up at levels of about five parts per trillion.[12] That's right, FIVE PARTS PER TRILLION! Pretty cool, hey? (By comparison, hydrogen sulphide – rotten egg gas – can only be detected in parts per billion.)

Petrichor, the wonderful, soul-lifting smell you get after rain on earth, is thanks in large part to the metabolites of soil life. To geosmin.

We've evolved around living earth, and our ability to sense geosmin could well explain why it feels so invigorating being in a garden, or so life-affirming to visit an ancient forest, as well as why the smell of summer rain lifts the spirits beyond the mere relief of it watering the garden.

Soil microbes make rain smell joyful.

Research that came out in 2020 suggests that soil microbes release geosmin probably not in an attempt to make us happy (though it's a nice thought), but rather to attract springtails,[13] tiny little insect-like soil dwellers that are gifted with the ability to leap. It seems a strange strategy, because springtails eat actinomycetes. But, there is an upside. The bacteria get a benefit, too: the springtails unknowingly carry the bacterial spores far and wide, ensuring the actinomycetes' continued survival. The bacteria release geosmin as they die, so it's not like they're cutting their life short, but rather seeding the new generation. It's a symbiotic relationship that is estimated to be 500 million years old.[14]

It seems like soil life is everywhere, doing everything. And that's not far from the truth. Living soil has far-reaching effects we are only just beginning to understand, from the microscopic level up to the macro-ecological.

To affect big ecosystems, however, soil life needs to become airborne. We now know that rain can cause bacteria to be lifted up and out of soil. Somewhere between 1 per cent and 25 per cent of bacteria that are released from soil can be put into the atmosphere by rain.[15] Even light rain. Gentle drops of rain can cause aerosols – tiny particles of moisture and microbes that defy gravity and can float on the lightest breeze. They can also float up into the clouds.

And this, it turns out, is really, really important in the production of rain.

Forty years ago, however, that idea was preposterous.

In 1978, certain wheat fields in the US state of Montana were declared free of a plant pathogen, *Pseudomonas syringae*. And so were the soils. In an effort to avoid bacterial leaf blight, which can reduce yields by half, seeds free of *P. syringae* were sown. And yet, despite clean seeds and clean soil, bacterial leaf blight still infected the wheat. Tests showed that *P. syringae*, the only known cause of the disease, had arrived on the plants from somewhere – and on the whole field at once, not starting in one corner or another.

Where, oh where, did the bacterium come from?

David Sands, a plant pathologist, looked to the skies for the answer. After the wheat fields were infected, Sands took to the air, flying up to 2.5 kilometres (1½ miles) above the fields, and using a low-tech method to test the atmosphere for microbes. He thrust his hand out of the window of the plane, clutching a petri dish, a sterile agar plate, and let microbes land on it. He even got a bit of frostbite in the process.

Back on Earth, the petri dish was incubated at 27°C (80°F) for 48 hours, and lo and behold, *P. syringae* bred up in all its glory. Not only had Sands found the leaf blight culprit, he'd also shown that bacteria can become airborne.

Sands suggested a novel theory in his research paper on the topic in 1982.[16] He suspected that the vector to bring the bacterium back down to Earth was rain. He knew *P. syringae* could cause water to freeze at temperatures that it wouldn't normally, triggering frost damage to leaves – the trick by which the bacterium can enter the leaf and attack it. Sands wondered if the bacterium itself was the genesis of rain in the clouds?

A flurry of movement in this area of science … didn't happen. Bacteria helping to form rain seemed too fanciful to be true. We were too busy building space shuttles.

It was two decades before Sands' hypothesis started to gain traction. And it wasn't until 2010, when researcher Alexander Michaud ran around collecting giant hailstones from the grounds at his workplace, Montana State University, that there was another vital discovery.[17] When Michaud dissected the hailstones, he found large amounts of living bacteria at their core.

Scientists looking at hail from all around the globe then found the same thing. They discovered hail often has a high microbial count, more so than you'd think from merely passing through air that has microbes floating in it. They also found our old friend, *P. syringae,* in the heart of hailstones. Since then, scientists have discovered microbes in snow and hail in South Africa, France, Australia and even in Antarctica. In fact, it's estimated that up to 80 per cent of all rain has bacteria in its core.[18]

For a long time, it was thought that the nuclei for rain and snow was inert. For water to condense and freeze, it needs a nucleus, a point to crystallise on. When water vapour sits in clouds, it can form little ice crystals around minerals, dust particles and the like. It needs to do this to actually rain. What scientists weren't prepared for was that living microbes not only act as nuclei, but they are better at it than minerals and dust.

Yes, one of the most amazing things that soil and plant bacteria can do is make rain. While it may appear logical that rain can disturb soil microbes and make them airborne – that rain releases bacteria, essentially – how can the opposite be true?

And why would it happen more often on bacteria than on specks of dust?

It turns out that, yes, ice can form around any nuclei, but for rain to form at higher temperatures (above –7°C/19°F and close to 0°C/32°F) – and to form more *easily* in many circumstances – living bacteria work better. Bacteria are the most active ice-forming nuclei in nature. It's estimated that between 70 per cent and 100 per cent of ice that is formed above –7°C (19°F) is on nuclei that are biological in nature – mostly living microbes.[19]

Sands' work was just the beginning. Scientists have now found that bacteria, fungal spores, diatoms and even algae can persist in clouds.[20] They often are the genesis of precipitation – rain, snow or hail. This discovery has spawned a whole new field (or sky?) of research, named 'bioprecipitation'.

These days, we actually breed P. syringae and use it in snow machines, but in a denatured form, so it doesn't cause plant disease. In nature, it's far from a single bacterium that does the work. Plant bacteria and soil bacteria seem to share the stage in terms of forming rain nuclei. In one study, from a single hailstorm in 2009 in Ljubljana, Slovenia, they found a whopping 1800 species of bacteria trapped by hailstones.[21] Soil and plant microbes have created this magnificent biofeedback loop, where they provide the nuclei for rain, while at the same time, the bacteria are able to move long distances in the clouds to colonise new plants and soils.

It seems that soils starved of life could be affecting rainfall patterns. It's early days, but that's certainly where the research is headed. We'd be mad not to consider the implications. We know that landscapes affect rainfall, when water vapour hits a landmass or mountain and condenses. And science has now shown that biologically active land-scapes can also dramatically affect rainfall. A forest can help generate rain. In fact, the Amazon rainforest generates a third of its own rain.

How much of that is because of the increased water vapour from the plants, and how much is from the presence of microbes for the water vapour to crystallise around, has yet to be determined.

If you have no living plants, if you plough earth, if you reduce its biodiversity and microbial populations, then you reduce the likelihood that bacteria can enter the atmosphere at the numbers required to trigger precipitation. Forget your rain dances, your cloud seeding using silver nitrate, your thoughts and prayers for drought-stricken landscapes. Bacterial changes could well be the culprit, and the saviour.

Chances are that biologically stunted areas – as much of our agricultural land has become – are less likely to provide the nuclei for rain. The more we bugger up our soil, the less rain we'll have in places that need it most. The good news is that more biologically active soil, healthy soil, is within our grasp. From the macro to the micro, what lies beneath matters. Healthy soil can not only make differences to big picture rainfall patterns, but also to us as individuals in ways we're only just beginning to understand.

CHAPTER 6

Look After the Soil, and the Plants Look After Us

As we've seen, soil is miraculous. It sustains life, forms rain, it elicits hormones that make us happy, and is the genesis of many modern medicines. Soil can gift us health.

That soil is a giver of life shouldn't need to be said. But as we've seen, and we'll continue to see, what we've done to soil in the last century has left us worse off.

We know one of the worst things we can do to our bodies, in the long term, is to eat the wrong things. Most of the diseases of the developed world are those of consumption. Over consumption and poor consumption of food can contribute to obesity, diabetes, heart disease, stroke, osteoporosis and a range of cancers. It's been estimated that it will cost US$30 trillion to deal with all of the world's diet-related diseases between 2011 and 2030.[1]

It's not just the developed world, of course, that is a victim of diet. Global hunger is a massive problem, and so are nutrient deficiencies, even in those who get enough calories to eat. While about a billion people are chronically malnourished, over 3 billion people are micronutrient deficient. We're talking major nutrients, such as iron, vitamin A, zinc and iodine.

Iron deficiency is huge, globally. Over 2 billion people are iron deficient, making it the largest nutritional deficiency on the planet, leading to about 20 per cent of maternal deaths.[2]

Vitamin A deficiency affects about 250 million preschool-aged children, claiming the lives of about 670,000 kids a year.[3]

Never heard of zinc deficiency, though? Well, neither had I, except in vague, academic terms. According to one report I read:

> **An estimated one-third of the world population is at risk for zinc deficiency; this deficiency is the fifth most important risk factor for diseases in developing countries. Worldwide, it is estimated that 800,000 people die every year from zinc deficiency ... comparable to the global mortality from malaria.[4]**

Comparable to the death rate from malaria! Yikes. Zinc deficiency – if it doesn't end up with you dead – leads to physical and mental retardation, delayed or decreased fertility, and a drop in immunity.

So we know that poor nutrition is affecting us, not only in developing countries, but also in developed countries. But what has that got to do with soil? Can we pin down a lack of nutrients to problems in soil?

Let's take a look at a single nutrient and see where the problem lies. Many of the world's soils are zinc deficient, and there's a general correlation between areas with zinc-deficient soils and those with zinc-deficient humans. Research on a more detailed level, however, shows that it's not necessarily the level of zinc in soil, but rather, whether any zinc that *is* in soil is bioavailable[5] – whether it is available to the plant, and to the human who eats that plant, or the human who eats the animal that has eaten that plant.

In other words, zinc can be present in the soil, but it's just not getting into the food.

Iron deficiency, which affects wealthy as well as poorer nations, is another useful case study. Very few soils are deficient in iron – but many people are. Part of this is due to differing diets, given that iron in meat is about 10–20 times more absorbable than iron in plant foods. However, the relationships between diet, individual differences (age, gender, etc.), and iron absorption are complex. To test whether the iron in soil affects iron deficiency in humans, you'd have to find a population of similar genetic stock (age, race, gender), eating a similar diet, but eating food grown on differing soils.

Luckily, such a scenario exists. There are a few regions of India, including Karnataka, where the soil is iron deficient, and where the rates of vegetarianism (and the diet as a whole) are comparable with neighbouring regions that have plenty of iron in their soil. Researchers didn't find higher rates of anaemia (caused by iron deficiency) in Karnataka's population. In fact, anaemia rates are actually lower in Karnataka than neighbouring areas with multiple times more iron available in the soil.[6]

So, iron deficiency is not caused by the mineral content of soil, or the overall choice of diet. What else could it be, then?

Could it be that the plants grown in iron-rich soil just can't access the iron, despite it being present? Nobody has yet done the science.

Why not? Because trying to measure nutrients within soil, within crops, within a diet, within a culture, and within a geographical limit is extraordinarily complicated. Depending on which type of rice you eat, if you eat 200 grams in a day, that 200 grams of rice could provide you with less than 25 per cent of your recommended daily intake of protein – or more than 65 per cent.[7] That's just the difference between rice varieties. Of a macronutrient.

One cultivar of apple can have 100 times more micronutrients than another variety,[8] before you even start to consider soil health.

Different varieties of apricot can provide less than 1 per cent, or more than 200 per cent, of the recommended daily intake of vitamin A.[9]

Then we need to look beyond differences in variety, too.

Nutrition is a very young science. Researchers only isolated the first vitamin, thiamine (vitamin B1), less than a century ago, in 1926. They only isolated vitamin C in 1936. Pretty much all the research up until the 1950s was on working out vitamins, and focusing on single nutrient deficiencies. Then we started to unpick fats and sugars – an issue we still seem to be grappling with after some dodgy science about saturated fats made political headway in the 1970s. In the first half of the 1900s, we probably knew more about plant nutrition than human nutrition.

In my lifetime, nutritional science has said lots of things. Eat low fat. No, eat less sugar. Don't eat eggs. It's okay to eat eggs. Meat causes cancer. Red meat is the best source of iron.

Nutritionists, when they're not telling us to eat margarine (what were they thinking?) can sound inconsistent. But one truth they do keep recommending is that we should eat a fair swag of fruit and vegetables. This is roundly backed up by science, because vegetables, in particular, contain some pretty good health-enhancing properties. Higher vegetable consumption is associated with better health outcomes, including heart health and a decrease in some cancers. We know that the indigestible starch – the fibre from plants – does wonders in feeding many of the bacteria in our gut. When those bacteria are well fed, they thrive, producing some pretty cool health-promoting metabolites such as short-chain fatty acids. The microbes also increase in number, and better ones take over, making your gut a much better place for the absorption of everything that isn't fibre.

Research also shows that if you eat 30 or more different plant products a week (and some of this can be as minor as a fresh herb in a dish), you're less likely to have autoimmune disease than people who only eat 10 a week.[10] So, we know that eating lots of veg – of any kind, really, but with some variety at least – is probably a good thing. Yet very few people in modern nations seem to be able to manage it.

But what if all those recommendations about what to eat – the 'five a day' and 'eat a rainbow' vegetable-focused messages – were failing because of one thing?

What if I told you that the reason you don't want to eat your greens is because of soil? That you don't eat them because they don't taste good – and that their taste is influenced, strongly, by the soil they are grown in?

We'll circle back here in a minute. First, let's look at food, and what's actually in our food.

Food is way more than the fats, proteins and carbohydrates it contains. It's also way more than the list of additives on the back of a breakfast cereal pack: riboflavin, niacin and folate. It's more than just getting the recommended daily intake of fibre. What's in our food is not just zinc and iron and magnesium and calcium.

Thinking of food as macronutrients (protein, fat, carbohydrate) and obvious micronutrients (minerals and vitamins) is about where the science stops. But such thinking is very 1920s.

Food is full of loads of other things that impart its particular look and feel and taste – an array of complex esters and antioxidants, phenols, phytochemicals (plant chemicals, such as carotenoids and flavonoids). These are all possible nutrients, too, meaning they can do us good. And they are all just starting to be studied.

Let's look at a few things we do know have been going on with some common micronutrients in our food, historically speaking.

A good place to start is with wheat, one of the three crops that dominate our diet (along with corn and rice). According to a paper from the *American Society of Horticultural Science*, looking at the historical nutrient density of wheat:

> **The reported average rate of decline of the six minerals (iron, copper, sulphur, phosphorus, selenium and zinc) is 0.20% to 0.33% per year, equivalent to 100-year declines of 22% to 39%, respectively.**[11]

The wheat we eat today is 22–39 per cent less nutritious than it was 100 years ago.

When I started on this book, I thought it'd be easy to prove or disprove a link between healthy soil and nutrient-dense food. Turns out I was wrong.

It's a remarkably hard ask to find data that backs up the link between living soil and nutrition. While nutritional science is relatively new, the research into what soil actually is doing for plants, and how plants can access nutrients, is an equally fresh endeavour. So far, both fields of study have been full of false starts. The cholesterol debate comes to mind. The advice when I was younger was don't eat more than two eggs a week, to avoid excess cholesterol intake. In those days we were told to avoid dietary cholesterol – when in fact the cholesterol you put in your mouth has nothing to do with actual blood serum cholesterol.

Now, we're about to unpick some of the links between soil, plants, diet and health. It can get a bit complicated, but bear with me. Here, I am going to wend my way through a maze of information to show that nothing we do – not the food we eat, or the soil we tend – is disconnected.

What I do find amazingly interesting, if you look back at what humans are designed to eat, is that different cultures managed to thrive on markedly varied diets. Some were virtually carnivorous, such as those living above the Arctic Circle. Some were vegetarian or vegan, at least in principle (even if a lot of insects and hidden animal protein used to sneak into their diets). And most cultures were omnivorous. Diets varied incredibly widely between the tropics and closer to the poles, from one season to the next, and from one year to the next. The human body is designed to eat a lot of different things, over different times, in different parts of the globe.

The clue, however, is in that very diversity. People can do what they like with their diet, but really, globally, the best advice is the most boring. Eat a wide variety of plants, mostly. Eat your veg. Sure, have some meat, some sugary treats, some fast food if you want – but if you eat well 90 per cent of the time, that other 10 per cent isn't really so important.

This is where soil comes into play. The actual benefit from that 90 per cent is driven by how that food was grown, and how that farmer or gardener treated their soil.

Nutritional science, like soil science, is playing catch-up with the complexity of whole systems.

The truth is, modern industrial agriculture means we are eating less nutrient-dense food than our grandparents did. This is not just with our wheat, as we saw above. And it's not only that we get most of our calories from about three species of grain. Or that most of what we eat comes from only 150 species of edible plant, rather than the 30,000 that are available. Or that we only use about five major animal species for 95 per cent of the world's meat.[12]

That lack of diversity matters, and it really matters in a farming and ecological sense. Eating the things we favour now – the 30 species of plant that provide about 95 per cent of our calories[13] – we're not getting the nutrition our bodies were designed for, and we're not getting the nutrient density we deserve.

For a start, the ratio of things we eat has changed even relatively recently. According to a 2017 article in *Trends in Plant Science*, looking at sustainable food systems:

> Between 1961 and 2013, the land area planted with wheat, rice, and maize increased from 66% to 79% of all cereals, while the land area planted with other cereals such as barley, millet, oats, rye, and sorghum – which have higher nutrient content – declined from 33% to 19%. As a result, the energy density of the cereal supply remained constant between 1961 and 2011, but the protein, iron, and zinc contents in the global cereal supply declined by 4%, 19%, and 5% respectively, with an overall decline of the nutrient-to-calories ratio.[14]

So, just by changing the crops we grow, we're getting the same calories, but fewer nutrients – there's 19 per cent less iron available in the global food scene. Hello anaemia.

That modern vegetables and fruits are lower in nutrients is fairly well documented. One of the first studies to make this clear was by Anne-Marie Mayer. In 1997, Mayer looked at results from the UK's *Chemical Composition of Foods*, and their analysis of mineral content from 1936 and in the 1980s.[15] Her study examined 20 fruits and 20 vegetables that were raw and peeled.

The only mineral that showed no real change in availability over the 50 years was phosphorus. All other minerals showed a substantial drop.

So, food *is* less nutritious these days, and it is thought there are three possible reasons for this. The first is the dilution effect; the second is handling (storage and transport); and the third is soil.

The 'dilution effect' is where the same plant absorbs the same amount of minerals, but because the plant is bigger, the minerals are more dilute. Less minerals per gram of food. From our look at wheat earlier, for instance, you'd have to eat more bread to get the same nutrition; more carbohydrate to get the same protein, iron, zinc and copper.

In vegetables, Mayer found statistically significant reductions in calcium (by 19%), magnesium (35%), copper (a massive 81%) and sodium (43%). In fruit, there was a drop in magnesium (11%), iron (32%), copper (36%) and potassium (20%).

She also notes that in fruit there was a significant increase in water, and a decrease in dry matter – essentially a dilution effect as well. (It's in the grower's interest to sell water for the price of fruit.)

Mayer admits there could be some sampling or researching errors in her study, but nevertheless does conclude there is a problem with the nutrient density of food. She poses a question that is still hard to answer, two and a half decades later: if we have reduced food's nutrient density, what does that mean for human health?

One study is not a trend, or reliable. But Mayer isn't alone in her findings. A widely shared 2004 study led by Donald Davis looked at the change in 13 major nutrients of 43 foods in the United States between 1950 and 1999.[16] He found that about half the nutrients had declined significantly, and around half had stayed about the same. Declines were seen in protein, calcium, phosphorus, iron, vitamin B2 (riboflavin) and vitamin C. Davis attributes the drop to changing particular varieties of vegetables and fruits in favour of less nutritious but higher-yielding ones.

A comparable study from the United Kingdom showed that carrots now have a third of the vitamin C they did in the 1960s, magnesium has dropped nearly in half, and sugars have increased by a third.[17] Copper also hovers at about a quarter to a third of 1940s levels. Interestingly, the study authors say this drop isn't significant, because you can still get your daily allowance of nutrients from other sources. Don't worry about dilution. Don't worry that pretty much every study shows a drop in nutrient density over time. Just eat more veg.

Encouraging us to eat vegetables *is* worthy for better health outcomes – but on its own, it's a flawed strategy.

Besides wheat, it's surprising how little work has actually been done on nutrient density in most of our food, especially vegetables like potatoes, say, or broad beans, or cabbage. Broccoli is an exception. It has been researched heavily because it's a popular vegetable, it has been bred extensively to give bigger yields, and it's interesting because the calcium in broccoli is as bioavailable as it is from milk – a rarity in the vegetable world. It also has quite a few phytochemicals that are believed to be important factors in heart health.

So what does research into broccoli show? Well, for a start, the amount of calcium in broccoli has dropped from 13 mg/gram in 1950, to 4.4 mg/gram in 2012.[18] So, if you don't eat dairy or fish or meat, your best source of calcium may be broccoli – and you have to eat three times as much of it today as you did in 1950 to get the same amount of calcium.

In 2009, the *American Society of Horticultural Science* reported that 'side-by-side plantings of low- and high-yield cultivars of broccoli and grains found consistently negative correlations between yield and concentrations of minerals and protein'.[19] The bigger the heads of broccoli, or the more tonnes of grain per hectare, the lower the nutrient density.

The dilution effect is very real, then.[20] And in large part it can be explained by cultivar selection. We've chosen higher-yielding plants, not more nutrient-dense plants; more carbohydrate per gram of food produced, less of the other nutrients. Pretty much all plant selection, for food, has been based not on nutrient density, but on yield and looks, and the ability to thrive in a high artificial-fertiliser environment.

Many research papers that discuss the drop in nutrient density in food point to dilution. Some argue these drops are insignificant, because in many ways we – in the developed world at least – can still meet our dietary needs through the food we eat: just eat more food.

However, modern transport and storage methods also deplete the nutrients in some of our food. A report in the journal *Food Chemistry* showed that broccoli can lose 70 per cent of its vitamin C and beta-carotene, and half its antioxidant activity, in six days after harvest. After being refrigerated, trucked, and put on a shelf for a couple of days in a shop, broccoli could also have lost 80 per cent of its glucosinolates[21] – a group of micronutrients widely regarded as very beneficial for heart health. They're also thought to be useful aids for cancer prevention. Another group of health-promoting micronutrients, flavonoids, had dropped by about 60 per cent after cold storage and time in a retail environment.

The big question for me is this: does soil health ever equal higher nutrient density? And if so, what does that mean for human health?

Research into organic growing methods is of interest here, given its increased focus on soil health, through the use of natural fertilisers, at the very least. A study in India showed that taro crops from organic plots had higher levels of dry matter, starch, sugars, phosphorus, potassium, calcium and magnesium than conventionally grown crops.[22] The soil also showed significantly higher (and bioavailable) levels of phosphorus, and higher soil organic carbon, exchangeable calcium, magnesium, iron, manganese, zinc and copper. The yield under organic management was also higher – by a not inconsiderable 29 per cent. In other words, more nutrient-dense food, in more biologically active soil.

Studies from the early 2000s also suggest it's likely that organic growing systems can increase the amount of phytochemicals in food. Writing in a 2006 edition of the *American Society for Horticultural Science*, lead researcher Xin Zhao says:

> The evidence overall seems in favor of enhancement of phytochemical content in organically grown produce, but there has been little systematic study of the factors that may contribute to increased phytochemical content in organic crops.[23]

Over a decade later, that's pretty much still the case.

In 1999, in an article entitled 'Mapping soil micronutrients', Jeffrey White and Robert Zasoski foreshadow the problem we would have finding an indisputable soil/nutrient density connection on historical data alone:

> Soils vary widely in their micronutrient content and in their ability to supply micronutrients in quantities sufficient for optimal crop growth. Soils deficient in their ability to supply micronutrients to crops are alarmingly widespread across the globe, and this problem is aggravated by the fact that many modern cultivars of major crops are highly sensitive to low micronutrient levels. Original geologic substrate and subsequent geochemical and pedogenic regimes determine total levels of micronutrients in soils. Total levels are rarely indicative of plant availability, however, because availability depends on soil pH, organic matter content, adsorptive surfaces, and other physical, chemical, and biological conditions in the rhizosphere.[24]

Phew. Let me decode that last bit.

White and Zasoski point out that even if soil mineral content is poor, biology – soil life – can fill the gap. Plants can still access a mesmerising variety of nutrients in living soil, even if that soil is low in some minerals. It's important that we choose a plant that can access those nutrients, and seek to find out what is actually happening in the rhizosphere – that wonderful interface where plants meet soil, and most soil life abounds.

Not all the science, however, shows that soil, or at least the growing method, influences micronutrients.

In a study from 2014, a team of researchers in the United States grew 23 different cultivars of broccoli, using conventional and organic methods.[25] They measured nine beneficial micronutrients with tongue-twister names like glucoraphanin, glucobrassicin, tocopherols and zeaxanthin. In other words, a modern micronutrient density study. And they found no overall difference in nutrient density based on farming system. The biggest difference was in the variety of broccoli (cultivars from the 1970s were more nutrient dense); there were also differences over the seasons.

The study only grew broccoli that was designed to be grown using artificial fertiliser and pesticides (conventional agriculture), so the argument could be made that the plants were perhaps unable to thrive in living soil. It was also a relatively short-term study. Still, it only showed a drop of phytochemicals because of the kind of broccoli we've chosen, the cultivars.

Longer-term studies show different results, however. A ten-year study on organic tomatoes[26] found that levels of the antioxidant quercetin (which can help relieve hay fever and hives) had increased 79 per cent, and concentrations of the anti-inflammatory kaempferol (believed to help prevent chronic diseases) were 97 per cent higher than in conventionally grown tomatoes. Results improved for every year the tomatoes were in an organic system, but didn't improve at all in the conventional one.

Spinach grown organically has also been shown to house more flavonoids and vitamin C, and lower nitrates, than spinach grown conventionally.[27]

It's also been shown, more generally, that adding artificial nitrogen to plants decreases beneficial nutrients in some crops such as lettuce, rocket, carrots and cabbages, and increases nitrates/nitrites, which are known carcinogens. In one study, there was 180 times more nitrate in butterhead lettuces grown in a high-nitrogen environment than in lettuces grown with lower artificial nitrogen.[28]

Something's definitely going on, underground.

More bioactive soil, microbially abundant soil, increases the nutrient density of plants, and can also increase yield. It's now been shown that by adding bacteria to soil, you can grow more wheat – and wheat that is higher in nutrients. In a 2017 study from India, researchers isolated three bacteria from the rhizosphere of plants, then bred them and poured them on the roots of wheat grown on the Indo-Gangetic Plain.[29] Now, we're talking three species of bacteria, out of an estimated 10,000 species in a handful of soil, so we're not really mimicking the true complexity of healthy soil. But even then, the researchers noted increased levels of nitrogen, phosphorus, copper, zinc, magnesium and iron in the wheat. The use of the bacteria also increased plant height, and yield.

Science is finding all kinds of effects that the underground economy can have in terms of how living soil can increase the nutrient density of our food. Inoculating soil with microbes has been shown to lift the micronutrient content of the roots, leaves and fruit of food plants. A bacterium, *Bacillus licheniformis*, added to tomatoes increases the flavonoid content of the fruit.[30] Adding specific fungi to the ground where legumes such as beans are grown has been shown to increase the saponin content in the resulting crop.[31] (Saponins are believed to lower plasma cholesterol levels, work as antioxidants and anti-cancer agents, and have a protective effect on the liver.) Soil bacteria added to the roots of rice have increased the amount of zinc and selenium in the harvested grain.[32]

How does this work? Well, new research is lifting the lid on what some people have been surmising for a long time. Life begets life.

At the root level, as we've seen, a plant is only able to access the vast majority of its nutrients through the action of microbes. The roots of a plant can access up to 2000 times more soil – and hence its minerals and multitude of chemical substrates – through the action of fungi, mites, worms, algae, protists and bacteria.[33] Living, healthy, vibrant, microbially abundant soil, that is.

We know there can be over 30,000 species of microbe around the roots of a healthy plant.[34] Sure, not all of the thousands of microbes in the rhizosphere are good for the plant, but the vast, vast majority are benign or helpful. It could be, if we want to grow specific crops, that in some cases we may need to nurture the more beneficial bacteria in the soil, just as we're learning to do with our guts.

We know the basics already. Just as eating fibre is good for the human gut and its resident microbiome, more subterranean life is generally a good thing for plant health – and for human nutrition.

Most changes to crops have benefited producers, very few the eaters. Cultivars that respond well to high nitrogen and don't require a complex underground ecosystem have all been developed to benefit those who aren't eating the harvest.

A farmer gets paid for crops based on a few things. Most important is weight: the bigger the better, so crops are selected for that. Fruit and vegetables must also look good, and be able to be transported and stored without losing too much visible condition. And growers have to compete on price.

Nowhere, in our current agricultural system, is there any motivation to grow nutrient-dense food. That's a major reason why we've gone down the dilution/poor cultivar/lifeless soil pathway for so long. Hence our recent success in growing the least nutrient-dense food in human history.

But times are changing. People are starting to look into how we can measure nutrient density more easily. An American mob called the Bionutrient Food Association[35] are trying to invent a simple, affordable way for people at home to measure the nutritional value of food beyond the old vitamins, fat, minerals way. In the meantime, they and others recommend using a refractometer to get an idea of what's in food.

A refractometer measures the way light bends as it passes through a liquid. The more it bends, the more dissolved things there are in the liquid. It can be used to measure sugar in solutions, salt in brines, and is a useful but indiscriminate indicator of what a plant contains. It can't show *specific* densities of individual nutrients, but it can show the *overall* nutrient density of a vegetable or fruit. You take a drop of juice from a fruit or vegetable, put it into a little window on the refractometer, and look through one end to see how the light has bent as it passes through the liquid. You won't be able to tell how much vitamin A is in your home-grown carrot compared to the store-bought one, for instance – but the more the light bends, the more nutrient-dense the vegie will be. That way you can compare carrots from the supermarket with home-grown carrots, organic spinach with conventionally grown spinach.

How you measure nutrients – actual nutrients, not just generalities – is far harder. If you want to measure phosphorus, for instance, you could use the spectrophotometric molybdovanadate method, I guess. Or the gravimetric quinoline molybdate method. They aren't techniques that are easily accessible, or even very pronounceable.

As I said, it's complex. But it doesn't have to be complicated.

The good news is, we have better tools at hand.

CHAPTER 7

Nutritional Dark Matter

Like cooking, and like gardening, nutrition is a humbling pursuit. If you think you know it all, you're wrong.

But, of course, we like to think we can know it all, especially because nutrition is a concept that we are all very intimately connected to. Being about things we put into our bodies, we all crave some certainty. If you'd like certainty, however, then don't look into nutrition.

There are lots of things science may not know. Remarkably, one of those is the role of food in our bodies. I used to believe that we could only guess or appreciate about half of what a plant or animal does for us when consumed as food. It turns out I was wrong. Half is a bold overestimate. The role of 99 per cent of what is in food is still unknown.

The way nutritional science has worked, until very recently, is that we find a chemical in the food (it could be something as common as carbohydrate, or a vitamin), and work out its role in nutrition. At first we looked at minerals (things occurring in the ground) and vitamins (things made by other living creatures, mostly bacteria). Nowadays, there are a few more in the list.

About 150 chemicals in food have been analysed, and their roles quantified, when it comes to their role in human health. But that is just a tiny proportion of the total. There are actually over 26,000 defined, distinct biochemicals in our food, according to a report titled 'The unmapped chemical complexity of our diet', published in *Nature Food* in early 2020.[1] These are the chemicals that make up our diet, and yet we know so very little about what they do for us, and what they do *to* us.

The lead author of that report, Albert-László Barabási, has coined a term for this great swathe of distinct chemicals: 'nutritional dark matter'. He thinks it's akin to the unknown 85 per cent of material, the dark matter, that we know to be in deep space from its effects, but can't see in the usual ways. The 26,625 food chemical components in his report are those that are actually defined, but, like dark matter, they are invisible to nutritional science because they're not considered important enough. And 26,625 isn't all that's out there. There are even more chemicals in food that science hasn't identified yet. When I checked the FooDB database,[2] a freely available food composition site from Canada available online at foodb.ca (the one Barabási used), they had already listed 67,000 chemical compounds that were detected, expected and predicted in food, but not quantified. It's much more than is listed in the nutrition panel on the back of your cereal box, that's for sure.

Barabási uses garlic as a magnificent example of how blind we are to the rainforest-like complexity in our food. The U.S. Department of Agriculture, the USDA, which takes responsibility for databases of nutrients and dietary recommendations, quantifies 67 nutritional components in garlic.[3] Sounds like quite a lot, and way more than just carbohydrate, vitamin C and sulphur. Of these 67 chemical compounds, 37 can be linked to disease, either in a preventive way, or in a negative way – things like B vitamins, manganese, copper, selenium, calcium. Their nutritional effects can depend very much on amounts; a deficiency or excess can cause health problems such as cardiovascular issues, Parkinson's disease and type 2 diabetes.

But, it turns out that 67 is just a drop in the garlic soup. There are actually 2306 distinct chemicals that have been isolated from garlic – and Barabási found that 485 of these have therapeutic effects. One of them is allicin, a compound that medical science believes has a protective role in heart disease.

Even then, it's not just the individual chemicals in the foods that is important here. What's actually of more value in working out a food's nutritional contribution is how these compounds interact, and what secondary metabolites they create. When we eat food, its composition is changed, and new effects can occur.

This will get complicated, with some funny-sounding chemicals nobody except chemistry geeks and medical researchers have ever heard of, but stay with me for a bit, because it's essential to the way we understand what happens to food in our bodies.

For instance, a thing called trimethylamine N-oxide (TMAO) is a chemical that, when found in the blood of otherwise stable patients, appears to worsen heart disease.[4] Red meat doesn't contain it at all, though it does exist in milk and fish. However, the TMAO in milk and fish doesn't seem to be a major source of blood TMAO, because it gets metabolised by the body into something else – so, eating fish and drinking milk doesn't increase your body's TMAO. Eating red meat can, however, increase your TMAO, even though the meat itself doesn't contain it.

This anomaly arises from metabolites. Red meat contains choline and L-carnitine, which are metabolised by our gut microbes into a compound called TMA, which our liver then converts to TMAO. Too much of this in your blood, as Barabási points out, can quadruple your chance of death over five years if you already have heart disease. Red meat can cause this problem, without actually containing any TMAO.

In other words, secondary metabolites are just as important as the original molecules when it comes to food's effect on our bodies.

And it turns out that those 26,000 food chemicals that Barabási identified can become over 49,000 secondary metabolites in our bodies.[5] Things like polyphenols, which are considered beneficial for health.

What all these secondary metabolites do, and how they interact, isn't as easy as counting calories. Even if we knew what they did, you'd still have to take cultural variations, individual differences and natural disparities in food into account. And here's the rub. It's the whole diet, the whole chemical cocktail, that matters in terms of nutrition. Allicin, which we met in garlic, actually reduces the amount of TMA in the blood, potentially decreasing the risk of death for those who eat red meat with the pungent herb. To reduce your risk of heart attack or stroke, perhaps serving garlic sauce with your steak might help.

Phytochemicals, antioxidants, polyphenols. These complex-sounding biochemicals are only present in food in tiny amounts – but small doesn't mean unimportant. Up to 95 per cent of polyphenols can survive the journey through the gut to the lower intestine.[6] There, some are broken down into secondary metabolites. Polyphenols are pretty good at promoting beneficial bacteria in the gut, such as bifidobacteria and lactobacilli, which can produce antioxidants. Research suggests polyphenols can also increase the body's ability to produce natural anti-inflammatories, with the potential to help alleviate mental disorders such as depression.[7]

A 2019 literature review in *Biotechnology Reports* looks at a closely related group of chemicals called phenols (which include polyphenols), of which about 8000 are known in plants. The authors say phenols are:

> acknowledged as strong natural antioxidants having a key role in a wide range of biological and pharmacological properties such as anti-inflammatory, anticancer, antimicrobial, antiallergic, antiviral, antithrombotic, hepatoprotective (functions).[8]

These biochemicals of plant origin can inhibit the start and progression of cancer, help our bodies fight off viruses and bacteria, reduce allergic reactions, help prevent blood clots and are good for the liver. Not bad for something invisible to nutritional science.

If you go on a date with someone who has just watched a bunch of funny videos, they smell better than if they'd just watched a scary movie. Their scent after watching something funny is more likely to make you smile. After watching something scary, the same person's scent is more likely to make you apprehensive. Interesting, but hey, what does that matter?

It matters because for a long time humans were thought to have a relatively weak sense of smell. We were considered the poor cousin in the olfactory stakes because we couldn't distinguish many smells compared to other mammals, such as dogs or rodents. Or pigs, which are famous for being able to sniff out truffles. Historically, it was esti-mated that humans could detect about 10,000 different odours – far fewer than pigs and dogs. We were, it was thought, stinkers at smelling.

Turns out, that's a myth. Humans can, potentially, sense a trillion different smells, according to research by Andreas Keller, of Rocke-feller University's Laboratory of Neurogenetics and Behavior.[9] We have fewer smell receptors, compared to some other mammals, but a bigger capacity to process the information those receptors give us; a bigger brain. And what *we* smell differs from other animals. Dogs, for instance, are hopeless at smelling bananas (probably because it's not in their interest, evolutionarily, to smell fruit), while we are very good at it. Other research has shown that for some things, we have a better sense of smell than monkeys, bats, rodents and even, just marginally, pigs.[10] We can track a scent across grass like a dog does. We can sense an airborne particle 0.01 nanometre across if it hits the right patch in our nose.[11]

We can detect the smell added to propane as a warning odour, down to 9 parts per trillion.[12] That's three drops in an Olympic-size swimming pool. Some smells can be detected at less than 1 part per trillion[13] – seriously low concentrations – but that's exactly what our noses were designed to do.

And what does this have to do with soil?

We've seen how healthy, fecund soil makes myriad more nutri-ents and minerals available to plants – and how plants can take up more of those compounds from living soil, and turn them into more nutrient-dense food. And we know that most of the chemical compounds in food are still unquantified by science.

Well, what if I told you that the single best way to tell the nutrient density of food, way beyond the capabilities of the refractometer, or any chemical analysis we're currently capable of, is simply to smell it?

Smelling food is what we do when we eat it. As we chew, the aromatic esters and polyphenols from the food are picked up in our airways and the fragrance shoots up the back of our nasal passages. Our tastebuds can perceive salt, sour, sweet, bitter, spicy, cool (like menthol) and umami flavours, but the actual sense of taste is only a tiny fraction of our experience; the rest is from fragrance. Most of what we talk about as flavour is based on smell. A more delicious potato, a more complex peach, a satisfying loaf of bread, a piece of meat that has more 'flavour' – these all have more chemical compounds in them. They have more inherent, complex, delicious smells. And they are all more nutrient dense.

So that's why the lettuce I once had at the kitchen table of a grower in Canberra tasted *sooo* good. Probably more phenols. That's why the carrots we harvest from our market garden where we focus on soil health have more flavour than commercially grown ones. Every home gardener knows their food is more delicious, generally, than anything you can buy. It gets that from several things. From its true seasonality. From its freshness. From the varieties we grow, yes. But, it also gets it from soil.

As the soil has improved in our market garden, so has the flavour of our produce. Better soil equals more nutrient-dense food. Not in a 1950s science way of measuring minerals or vitamins. But we know all those phytochemicals and antioxidants are doing us good. We can smell them, or the biochemicals associated with them. We taste them in our food, and we know when something is nutrient dense or not.

Anyone who doubts the ability of the human nose to sense nutrient density probably doesn't drink wine. Or coffee. Coffee nerds are forever espousing where the beans were grown, and how they were harvested, stored, roasted and more. Every winemaker can tell you about their soil, and how it changes within one vineyard, and even within one row. Every great wine is an expression of place, and so is great food.

Let's stick with the wine example for a moment. Think of all plants as grapes, for a second. Grapes are all the one species, *Vitis vinifera*, but there are multiple varieties, or cultivars. Chardonnay, cabernet sauvignon, albarino, sangiovese – these are just a few of the varieties of the same species. People buy wine based on variety, and these do vary a lot, not just in colour, but also in flavour. But each variety also has clones. Pinot noir clones 828 and D5V12 are supposedly better suited to a vineyard near me than clone MV6.

So, the variety – or cultivar – of a plant can form a big basis of its taste. Thompson's seedless grapes taste different to pinot gris grapes, for instance. But, variety and clone don't completely determine the end result. Season does, and soil does. The exact same variety and clone in different soil, and in soil treated in different ways, has a different flavour. That's what winemakers trade on – the expression of flavour from soil and farming method. They call it *terroir* in French, an expression of place. Then, on top of terroir, each season (called 'vintage' in wine speak) – even within the same row and from the same vine – has a different expression in the glass. Why? In large part because of soil.

Why do we know so much about grape cultivars for wine, and not about cultivars of asparagus, or zucchini, or capsicums, and the role of soil in their expression? Because wine is worth a motza. It's way more valuable than a parsnip or a spud. And because we pay more for it, we concentrate more on it, sensing the array of chemical components that the grape provides. Swirling and sniffing a glass of wine from a $30 bottle of pinot noir means you're using your incredible natural gift, your nose, to tell you what it contains. You're smelling some of the trillion aromas that we're capable of sensing,[14] aware of more than just a fraction of the nutritional dark matter that science has yet to quantify or establish a role for.

And wine – fermented grape juice – is relatively simple compared to the complex array of nutritionally viable chemicals that can be in food.

What is true of grapes is true of all food. Flavour is a reflection of nutrient density, the best indicator we have of soil health when we ingest our food. Yes, cultivars matter. Not all cauliflower cultivars are created equal, and not all are good in every growing region. And seasons also have a strong influence on flavour, and hence nutrient density. That's hardly news to a home gardener, finding that a hot spell makes your radishes spicier, or a cool snap might lead to less aromatic strawberries. But the important thing to remember is that soil health is also vital for the best-tasting food, and the food that is best for you.

As we've seen: if you look after the soil, the plants will look after you.

Much of the complex new science is pointing to this thing that good growers have known forever. Something that makes sense, and that we as a species are clearly designed for. Our noses and our bodies are built to work in tandem to find, cherish and use the foods that provide us with the best nutritional benefit.

It's really not a long leap from soil health, to plant health, to human health.

We don't fully utilise our great gift of smell, but we can train it.

Some people are more in tune with their nose, and their food, than others, particularly those who taste for a living. In a novel study from 2013 in Italy, Luisa Torri led a team that looked at the flavour of bread and how it was affected by the wheat used.[15] They employed a tasting panel to test old varieties of wheat, and a modern variety that was grown with, and without, a particular fungal strain. The theory is, if fungi gives the wheat more or different micronutrients, you'd be able to taste the difference in the resultant bread. And you can.

The panel all had over two years sensory experience in tasting food and drink, and all did eight hours of training in how to assess bread specifically (visually, texturally, and in taste). They rated the bread made from historic wheat varieties really well on flavour. But they also rated the modern wheat, grown using the fungal application, very highly too, compared to the same wheat without the fungi. It was distinguishable based on taste, simply because it was grown with a different fungal network in the soil. (An electronic 'nose' could also tell the difference, and backed up the human tasters' experience.)

That a change in soil fungus can alter flavour has also been shown with tomatoes, lettuce, basil and more. More life, more variety below ground, can equal more flavour, and more nutrients per gram of food. Maybe not more carbohydrates, but more of the things that also matter, things that aren't in our vocabulary yet. The Dark Matter.

That we are gifted this sense of smell – an ability to detect thousands of chemical components in our food as we eat it – to seek out more complex, delicious food is no biological accident. It's what we're made for. Simplistic nutritional science and nutritional analysis of the food we grow has nothing on the hooter that pokes out the front of your face.

The reason most of us don't eat our veg is because it's boring. Modern cultivars have been stripped of nutritional value and flavour by a system geared to the grower, not the eater. We've focused on quantity not quality.

This doesn't bode well for our long-term health. If you grow food faster, food that is full of macronutrients and has less micronutrients, then you need more of it just to stay as healthy. The hollow foods we grow lack nutrients not only relative to their size, but also in comparison to what should be there, full stop. It's those missing nutrients that make food taste more of itself. Lack of nutrients could be why you don't like carrots – because you may have never tasted true carrotiness. Or have tasted a bland or bitter carrot, something out of balance with what it could, and should be.

In other words, poor soil can lead to nutritionally poorer dinner on the table, but importantly, for the chef in me, it also leads to gastronomically poorer meals.

I know I've harped on about it, but the fact is that nutrition is really, really complex. Food is really, really complex. Soil is really, really complex. Growing food is really, really complex. Their interactions can't yet be distilled into a workable formula for good crops and good health. But, the great thing is, eating doesn't have to be complicated.

A century of nutritional science has taught us that simple fixes just aren't fixes. To overcome an increase in heart disease, for instance, focusing on fat in our diets did virtually nothing to reduce rates of heart failure (though medical intervention did). We tried to blame fat without looking at whole diets, or the complexity of diets within a culture and landscape. What actually does work is variety. So a Mediterranean diet that targets a mix of foods, not a single 'super-food' or a single villainous component of food, works better on a range of health measures than simply a low-fat diet. Eating as many different, complex-tasting delicious plant foods we can, with a small amount of pasture-raised meat if you're that way inclined, is what we are constructed to eat and enjoy.

Of course, a lot of people on the planet don't have the luxury of endless variety in their diet. They aren't worried about a faulty immune system from depleted vegetables, and microbe-free food. They'd really like to get *enough* to eat for a start. That is an economic and political problem, more than anything, but what should concern us most is ensuring that the food those people *do* get is as nutrient dense as it can be. And that applies to disadvantaged people within richer nations, too.

All around the globe, we have focused too heavily on single nutrient deficiencies and efficiencies, and not enough on the whole diet. In developing nations, monoculture staple crops lacking in iron, vitamin A, zinc or other minerals aren't helping malnutrition, only exacerbating it.

How can we be sure that many of those 70,000 chemicals that are supposed to be in our food are making it to the kitchen? How can we get the most delicious food on the table, and how do we ensure that those who struggle with food security are getting the maximum number of nutrients in the food that they do manage to source?

By looking after the soil, that's how.

In ground-breaking research, scientists now know that we don't just 'taste' food in our mouths. We taste it all along our gut. Fractions of the food, and the microbial DNA, pass into the gut lining, and messages are passed directly to our brains.[16]

Until recently, we thought that all the food we ate was denatured in the act of digesting, including the DNA and RNA, the genetic codes from that food. It was assumed they were destroyed by gut processes. Wrong again.

Peer-reviewed papers in magazines such as *Cell Host & Microbe* are showing how tiny fragments of plant and bacterial DNA do get absorbed into cells in our gut.[17] They suggest it's very likely that the food we eat can affect the way our bodies sense the world, in ways that go far beyond just fuel.

This is mind-blowing stuff.

In one experiment, certain chemicals of a water-stressed plant – which are different to a non-stressed plant – were shown to pass into the cells of the human gut.[18] As they did so, they were passing on a message from the food to our gut – and ultimately our brain – about the state of the climate. Is it possible that our food can be telling us that the land is dry, the plants under stress? It seems so.

In other cutting-edge research, there's also evidence that phytochemicals and their metabolic products can act as prebiotics. They can actually change the gut's microbial composition, to inhibit harmful bacteria and stimulate the growth of beneficial bacteria.[19] This has really strong implications for our so-called 'gut–brain axis' – the deep connectivity between our food and our mental condition. It might sound far-fetched, but this is now accepted science.

L-ergothioneine, the anti-ageing antioxidant we met in Chapter 5, is only made by soil microbes, and is drawn up by plants. From there it enters our body, through our gut, and can be measured in our bloodstream. Recently, research has shown that it can actually cross the blood–brain barrier, which protects our brain from a whole bunch of chemicals and pathogens.[20]

In other words, soil microbes make L-ergothioneine, an anti-oxidant chemical that is passed to plants, enters us, and can survive the journey directly into our brain. From there, it is surmised, it can exert anti-dementia properties.

Soil really is our friend.

Our food, the very DNA of it, talks to us. It cues responses in our gut. Our gut signals to our brains. The brain and gut are constantly sending messages to each other. What we eat affects our mood. It affects our immune responses. Our food could well be involved in gene regulation, organ stimulation, and be a system engineer. If we strip our food of micronutrients, and a natural microbial community, we risk changing things we don't yet fully understand.

No wonder many researchers are calling the rhizosphere microbiome a 'super-organism', just like our own microbiome. We are not alone, and we don't act alone. Everything is connected. Every one of the possible 100,000 root exudates we met in Chapter 4 are biochemicals a plant uses to communicate, to protect itself, to nourish the underground ecosystem; they are nutritional dark matter that each plant produces in its structure. Maybe not 100,000 of them in all plants all the time – but they all exist, or can exist, in the meals we put on the table, and they all hold the possibility of having an impact on our health when we ingest them.

Our bodies are designed to consume a smorgasbord of naturally occurring biochemicals in our food. To ensure we're getting what we all need and deserve, we just have to go back to the source, to the origins of our food.

And that brings us back, once again, to soil, and just what is happening to this precious, vital resource.

CHAPTER 8

Here Today,
Gone Tomorrow

I'm standing in a vast erosion gully in Queensland, in Australia's north-east. This, the oldest continent on Earth, with the poorest soil, the lowest rainfall and the most desert of any land on the planet, is no stranger to eroding landscapes. But this gully, chiselled through a cattle-grazing station over a million acres in size, towers over my head. It's a dry creek bed where none previously existed, carved into the landscape by water when it rains – one of many that eventually dump into older, real creek beds, which then lead into rivers. Those rivers then feed into the Great Barrier Reef, which is imperilled in part through silt. We've just finished filming some land clearing for a documentary on meat production, and I've taken a moment of contemplation here to feel the scale of our national loss.

It's not the biggest erosion gully I saw from a helicopter ride over this land, but it dwarfs me nonetheless. I'm shaken and saddened. I feel a visceral sense of the loss it represents.

And what a loss it is. Australian agricultural land loses about 1.5 billion tonnes of topsoil a year as it moves downhill, thanks to water erosion alone.[1] That's about 2.2 tonnes per hectare, per year.[2] When you spread it out, it doesn't look like much; less than a millimetre a year. But over a 30-year period, you can lose 27 millimetres of topsoil.[3] Which again may not sound like much, until you consider that it takes about

1000 years for nature to build 10 millimetres of topsoil over much of our continent[4] – and we're losing it nearly 100 times faster than it can be made. The half-life of soil, the amount of time it will take to lose half of what little is left of our topsoil, is now measured in decades across most of the nation.[5]

If you loaded all the soil Australia loses each year to water erosion into railway cars, the train would stretch around the globe. Seven times.[6] And the entire loss of soil, including through wind and tillage erosion, is approximately double that.[7]

When you see a brown creek, or a brown plume at the mouth of a river as it heads into the ocean, or a dust storm – this is topsoil being lost in vast, irretrievable quantities. It's estimated that Australia has lost half its topsoil in just over 200 years since Europeans arrived.[8]

If there's one thing we know, it's that humans are damaging the Earth. We do it in all kinds of ways. We dig up things and leave massive scars. We overfish and deplete our oceans. We chop down forests and create deserts. We build cities where once we farmed. Wherever humans have been, we have had an impact.

When we first appeared on the planet, when our ancestors became humans, we found ourselves living in the original Garden of Eden. Plants dominated the landscape. Since humans came along, however, we've cut down, mown or paved over so much plant life that it's estimated we've lost half the total biomass of the world's plants.[9]

Globally, our modern urban areas take up enough space to be the world's seventh biggest country, larger in size than the entire European Union.[10] Houses, roads and skyscrapers where once all was green.

We've also done devastating things to our un-urbanised land, farming often being responsible for the worst of it. Fortunately, however, this also means that better farming can help repair the land we've ruined.

Before we get into that, a caveat. I don't want to play the blame game. This book isn't about pointing the finger and demonising those who came before. But it is about accepting what we've done wrong, to see what we can do better. Nobody set out to intentionally bugger up the world. In many ways we have done just that, though, through ignorance, self-interest and poor historical reference. What we have done, and got away with in the short term, we can't continue to do. It's not denigrating our forebears to acknowledge that the things we did to build civilisations, to pull people out of poverty, to feed entire nations, also had an impact that we can now too readily see. It doesn't mean we can't celebrate the past while at the same time acknowledging the failures.

Before white colonisation, Australia's erosion rates were many times lower than they are under modern agriculture. Around the Great Barrier Reef – a reef bigger than the whole of Italy – erosion rates are estimated to be five to ten times higher than prior to European colonisation.[11] And that change has happened rapidly: white settlement of northern Queensland began in earnest only around 150 years ago.

It's not just here, on this old, fragile continent, that soil has been squandered. The United States loses about the same amount of topsoil per year as water erodes in Australia: 1.5 billion tonnes, washed or blown away.[12]

When I first looked into erosion, the number bandied about was that we were losing an average of 2–3 tonnes of topsoil per person, per year, around the globe. That's a lot of soil, for a lot of people.

Traditionally, of course, if we ruined land, we'd just move to the next bit. Then the bit after that.

But I couldn't believe the 2–3 tonnes figure, so I looked up a bunch of references. Turns out I could find no estimate lower than 2 tonnes – and it's probably conservatively more like 6–7 tonnes per person, per year, of fertile, magical topsoil that is lost forever into our rivers and oceans.[13]

In fact, according to the Global Soil Partnership, led by the UN's Food and Agriculture Organization (FAO), a whopping 75 billion tonnes of soil is eroded every year from arable lands around the world – with an estimated financial loss of US$400 billion per year.[14] That's nearly 10 tonnes per person, or about 9 kilograms (20 pounds) of topsoil lost for every single meal we eat.[15] Nine kilograms for every breakfast, every lunch, every dinner, for every human on Earth.[16]

And that's just arable cropping land – not the soil lost by over-grazing of livestock, which is hardly insignificant, as I witnessed in that erosion gully in Queensland.

Australia has obvious soil erosion problems, but even relatively fertile continents lose soil. The European Union is losing 970 million tonnes of soil per year due to water erosion alone.[17] That's an amount that could cover an area twice the size of Belgium with 1 centimetre of soil.[18]

One centimetre of topsoil is all that much of Australia has – and that took 1000 years to accumulate.

And because that happens every year, we're compounding an already huge problem.

Jerry Glover, an agro-ecologist at the U.S. Agency for International Development, has calculated that globally, by 1991, an area of farming land bigger than the United States and Canada combined had been lost to soil erosion since the agricultural revolution.[19] That's a mere 12,000 years ago, a blip in geological time.

This erosion, of course, has incredible impacts on the ability of farmers to farm. Soil loss equals less food. In Africa, for instance, farm yields have dropped up to 40 per cent thanks to soil erosion in recent years.[20]

You'd think a problem this massive, this globally significant, would've been a hot topic for years. I remember seeing the deeply scarred and eroded paddocks in New South Wales, where I lived during the big drought of 1982–83, so visible erosion wasn't a secret.

But while some work was being done, particularly in continents like Australia where erosion strikes hard and fast, it's only in recent times that it has been taken seriously on a global scale. According to the FAO, 'there was more literature published on soil erosion in the three years between 2016 and 2018 than in all of the twentieth century'. I do not make this up: 7348 articles and research papers in three years, compared to 5698 over a century.[21]

The good news, of course, is that we *are* now starting to take it seriously. But are we too late? To answer this, first we need to understand what causes erosion.

ARABLE LAND: THE THIN EDGE OF THE APPLE WEDGE

To visualise how little of the Earth is available for agriculture, take an apple, cut it in half, and discard half. Cut it in half again, and discard another piece. What you're left with is a quarter of the apple; this is how much land there is on Earth, roughly, compared to oceans.

But not all land is good enough for growing food, so cut the remaining quarter in half again. One piece of this – in other words, one-eighth of the apple – is no good for living on or growing food. It's too cold, or too dry, rocky, alpine or swampy. So discard that piece, too.

Now you're left with the remaining one-eighth of the apple. Now, cut this into four equal-sized pieces. Three pieces represent where people can live, but where the land isn't necessarily good for growing vegetables and grains. It may be reserves, national parks, cities (on once arable land), or degraded agricultural land. It also includes grasslands that we can graze animals on, that isn't suitable for agriculture. So discard those three pieces, too.

The last piece remaining represents all the arable land on Earth available for farmers to grow food on. What you're holding represents about one-32nd of the Earth.

But in reality, the amount of arable land is even less than that.

The peel on that last remaining sliver of apple, thin as it is, is way, way thicker than the Earth's topsoil – the bit responsible for all the world's land-based growing. In fact, the skin is about the thickness of the Earth's crust, which is over 80,000 times thicker than the topsoil.

The topsoil is probably best represented on our apple by the flicker of light being reflected off its skin.

Put simply, bare earth is bad. Plants try to cover bare earth, and bare soil wants to be covered. Water and wind can do little to disturb the topsoil when there's a verdant mat of greenery protecting it.

In the big scheme of things, cutting down a clump of trees is the worst thing you can do for soil. A good old forest doesn't really suffer much erosion compared to open ground. And if you cut down a forest, you immediately speed up the rate at which water can impact soil. Like mini missiles, the water droplets from rain loosen soil with each strike, hitting the ground at up to 30 kilometres an hour (20 miles/hr).[22]

It's estimated that the energy of annual rainfall on a hectare of land is equivalent to the explosive force of 50 tonnes of TNT – which equates to about 20 tons of TNT per acre.[23]

In a forest, there's a buffer. A tree's leaves slow the raindrops down. Then there's the other things a tree provides, like leaf and bark litter on the ground to protect against raindrops, and drips from its own leaves and branches. Trees also provide stability to the soil through their roots.

Once the forest is gone, all the roots die, carbon is lost, and life diminishes in the soil for a while. The raindrops can damage any bare earth, and as they rapidly fall together in a storm or heavy rain, there's the rush of water downhill. This is the cause of what I saw standing in that deep scar in the heavily cleared, cattle-grazing country in Queensland. Water moves most things downhill, eventually.

The next worst thing when it comes to erosion, after cutting down an existing forest, is digging the soil. This exposes bare earth to the effects of rain, and to wind erosion. The problem is, growing food, particularly vegetables, often produces bare, dry earth. Just think of the paddocks where your broccoli grows, or your parsnips were dug.

Water erosion can happen in a forest, but far more slowly than in open country. Wind erosion doesn't happen in woodland; it can pretty much only happen when soil is exposed to the atmosphere and becomes dry enough to be airborne.

Wind erosion – a massive problem in a drought-affected land like Australia – is often called a dust storm. I call it a soil storm. Many times over the last century, the glaciers of New Zealand, over 2500 kilometres (1500 miles) to the east, have been stained with the red and black 'dust' of Australia's interior. This phenomenon, which can be clearly seen in the layers of those glaciers, rarely occurred prior to European settlement.

In the United States, massive dust storms stripped the fertility from much of the Midwest in the 1930s. This created the infamous Dust Bowl, which was so severe that it still informs some public policy today.

After water erosion, called alluvial erosion, and wind erosion, called aeolian erosion, comes what is known as colluvial (gravity-transported) erosion, which is much harder to see. But see it you can, if you think of the simple action of gravity – stuff moving downhill, but not through the actions of water.

A lot of this relatively unseen colluvial erosion is that of tillage erosion – which is the kind that occurs when you plough the land to turn the soil. Tillage erosion wasn't even identified by scientists until the 1990s, and is still poorly understood.

Tilling the soil is a losing game, not only because it upends earth ready for all three types of erosion. It also destroys soil's microbial home, and soil's major glue, glomalin, which we met in Chapter 4 and will come back to again.

Erosion can also come from overgrazing with livestock. It can also occur because of severe weather – which we're forecast to see more of.

Erosion works far faster than the glacial time frames it took to make the building blocks of our soil, and far faster than fungi and algae can break down rock into soil's component parts. We can, as in all things, destroy faster than we can build. All that sand, silt and clay that the last ice age gifted us? All the cracked stone and sediment that has built up on arable land? Well, we've been very good at washing and blowing it away.

We, as individuals, as nations, as a species, have a limited amount of soil. It takes a very long time to make, and is far easier to lose than we imagined.

Nobody a hundred years ago would have been wondering if we'd have enough land to grow the world's food by the end of the 21st century. Modern farming methods, including the use of farm machinery and an array of chemicals, have sped up our ability to ruin topsoil. In some places it happens in years. In others, decades.

But inevitably, we're losing the very thing that holds the key to life on land.

Soil is lost, currently, about 30–40 times faster than it is replaced globally.[24] It isn't hard to do the maths. It isn't sustainable.

Once the life has gone from soil, it is very easy for the physical soil to depart. And once you lose the structure of soil, and the life, it's only a short step to desertification, if rainfall is sporadic. Soil, like other ecosystems, reaches a tipping point, a moment when so much of it is degraded that it's almost impossible to get it back.

Currently it's estimated that 12 million hectares of land are lost to desertification, globally, each year.[25] Even great soil is being lost at astonishing rates. In Iowa, a pretty fertile part of the United States, topsoil depth more than halved from around 35–45 centimetres (14–18 inches) in 1900, to 15–20 centimetres (6–8 inches) a century later.[26] Because Iowans still have topsoil, the problem seems less drastic, but at this rate even Iowa could be an agricultural desert in another 100 years.

We humans have had soil in our control for a very long time. Often we've misread it, only using a single year, or the span of a single human lifetime, to try to assess our actions. These time frames, and visible changes, don't always reflect our impact on the substance under our feet.

The good news, however, is that while soil is fragile, soil can actually be made.

One way in which soil can be made has been known by home gardeners for over a century and a half. It's the thing they get smug about when they show you their patch. It's what Charles Darwin, who first espoused the evolutionary theory of survival of the fittest, spent his dotage considering.

It's worms, and their story will gladden your heart.

CHAPTER 9

Big Ones, Small Ones, Skinny Ones, Fat Ones: Worms

Graeme Stevenson is skipping around with glee in an almost Gollum-like prance. He's very, very excited by what he sees, and he wants me to share in the pleasure.

Graeme has come to our farm to look at poo. He drove five hours to peer underneath a few of the cow turds that are scattered around our property. He's let drop a few 'shits' (as in, 'that's a beaut shit', and 'what's under this shit?') – which doesn't bother me, but perturbs a film crew who are recording his visit. 'Shit' won't make for family-friendly viewing on a show we are making. But I don't notice the swearing as much as the jig Graeme is performing.

Graeme is a self-proclaimed 'poo-ologist'. He visits schools, tours farms, digs gardens looking for scarab beetles, a.k.a. dung beetles and worms. He's here at my farm to see if we have any decent dung beetles, because we have cows, and therefore we have dung.

Since 1967, around 50 species of dung beetle have been introduced into Australia, and 25 have become established.[1] All over the nation, soils have benefited from these turd tunnellers. Millions of dung beetles flit from pile to pile, nourishing themselves while nourishing the soil. Not only do they move the poo down into the earth, but they can aslo carry mites that attack fly larvae.

Cow dung, in the quantities we have from our small herd, isn't a toxic waste problem. It's fertiliser. But if the dung sits on the ground for too long, there's a risk that flies will lay their eggs in it, and we'll end up with more annoying blowflies that signify summer on the land. Dung beetles smell a freshly laid cow pat, fly in from up to a kilometre away, and then bury themselves in the muck before digging holes in the earth, small amounts of cow pat in hand. They bury poo underground.

Now, there are about 90 species of dung beetle in the United States, of both the native and introduced kind.[2] In Britain and Europe there are more than 60 species.[3] There are about 200 species of dung beetle that are native to Australia, too[4] – but they're adapted to the dry and fibrous pellets left by native animals. A big, wet cow pat is a different thing altogether, and many native beetles struggle to consume them.

It only takes a cursory look for Graeme to declare that I have a healthy population.

It's not, however, these magnificent recyclers that has Graeme dancing. It's a worm. *Aporrectodea longa*, the deep digger, an earthworm more commonly found in the warmer climes of northern Tasmania than in this part of the state.

A. longa tunnels up to a metre down, leaving soil more fertile, moving nutrients and making them more plant available. They're generally considered a heavy lifter in the worm stakes. Graeme is over the moon seeing one of them here. I've never seen anybody scampering about, deliriously happy, at the sight of a worm before. It gives me such joy to watch, but I'm not sure I can match Graeme's glee over a wriggly little thing I once thought I could eat.

While I like earthworms, I'd had a bit of a bad experience with one. A few years prior, I'd been told that tiger worms could be purged and would taste like abalone, the prized shellfish known as paua in New Zealand. They didn't. They tasted like a troll's armpit, an evil flavour borne from something close to the grave. Perhaps the species' name gives a clue as to its palatability: *Eisenia fetida*. And fetid it tasted. (Apparently even trout find them unpalatable, according to Australia's pre-eminent scientific authority, the CSIRO.[5])

Graeme, a man who has spent most of his adult life trying to get people interested in soil, isn't trying to get anybody to eat worms. Rather, he wants to encourage everybody to get them into their gardens and paddocks.

And he isn't the first to spend a lot of his time thinking about worms.

In 1881, a little book with a woodland-green cover was published in Britain. Far from shouting out the author's credentials, it seemed a modest tome, but was penned by the master of evolutionary theory, Charles Darwin. And while the front cover was blank, the spine was emblazoned with gilt lettering proclaiming 'Vegetable Mould and Earth Worms'. The full name on the title page was *The Formation of Vegetable Mould Through the Action of Worms, with Observations on Their Habits.*[6]

In this, his final publication, which he described to a friend as 'a small book of little moment', Darwin observed not the variation in turtles from the Galapagos Islands, or the fantastical marsupials of Australia, but mostly his home country, England. His small book on worms, and what they do to create vegetable mould – dark organic matter in healthy soil; humus in modern speak – quickly sold thousands of copies.

Darwin was writing of his last two decades of work, and from forty years of observing worms. Why on earth he'd chosen to write about these, of all the globe's creatures, remains a mystery. Early on in the book, he presciently encapsulates pretty much all that has been thought of soil science before and since: 'The subject may appear an insignificant one, but we shall see that it possesses some interest.'

Some interest? The master of understatement, Darwin obviously knew that in comparison with his earlier work, *Worms*, as his book would be quickly abbreviated, would garner less excitement than the discovery of a new star, or new lands. 'Oh, that's rather nice of Charles, isn't it? To write something about itsy-bitsy worms. A small book on worms for the gardener. Jolly good show, eh, what?'

But what Darwin observed underlies why he is so famed in evolutionary circles. This is no poetic, romanticised view of worms. It's no bland look at a seemingly innocuous, dreary, blind and deaf subterranean creature. It's hard science. He went back to basics, observed, measured, asked questions that needed an answer. One of his original inquiries came when he saw rocks disappearing into the earth in an unploughed field on his farm. What mysterious force was causing the rocks to sink? Darwin saw something that puzzled him, and set his not inconsiderable brain upon it.

So, Darwin started watching worms. And became bedazzled by them. He wondered why there were twice as many worms in a kitchen garden than in a cornfield. Why they were bountiful in forests. And he began to realise that worms, despite their small size, did an awful lot of work when it comes to cycling plant matter.

Darwin pinned leaves to the ground and watched worms nibble the edges and try to pull parts into their burrows overnight. He measured and described the types of leaves pulled into wormholes, how they were usually pulled in point first (the easy way). He measured how much soil each worm might bring up to the earth's surface during the night, and how much of this moved downhill.

But most fascinating for me in Darwin's book is this statement, especially in light of the current drastic erosion of soil:

Farmers in England are well aware that objects of all kinds, left on the surface of pasture-land, after a time disappear, or, as they say, work themselves downwards. How powdered lime, cinders, and heavy stones, can work down, and at the same rate, through the matted roots of a grass-covered surface, is a question which has probably never occurred to them.[7]

When I read that, being from a country that usually erodes soil, and having spent a childhood finding things unearthed by wind or rain rather than submerged by soil, I was gobsmacked. But for Darwin, that was normal. In the good soil of Britain, when you didn't plough, large rocks could disappear into the ground. And he suspected worms. If it proved so, he wanted to quantify it. Just how much work do worms do? They may be small – but if you have enough of them, each one doesn't have to do that much.

How many worms did Darwin think were living under the soil? One of his contemporaries, German physiologist Viktor Hensen (who, incidentally, coined the term *plankton*), had published his own work on worms a few years prior. Hensen estimated there were 133,000 earthworms per hectare (53,750 earthworms per acre), based on the number he found in gardens.[8] Darwin concurred.

One thing Darwin noted was that size and mass didn't seem to make any difference to the rate at which rocks 'sank'. The specific gravity of the stones didn't seem to affect the rate of disappearance. He wondered if it was due to the subversion of the soil as worms dug under rocks that caused them to sink, or whether worm castings – the fine dark matter that passes through a worm's guts – gradually caused soil levels to rise.

What worms do, Darwin postulated, was consume earth and organic matter, and make *vegetable mould*, humus, the holy grail of home gardeners. As worms digest and dig, they bring up other matter to the surface, moving earth around.

Darwin's small book actually wasn't insignificant in effect. It kicked off plenty of debate, and inspired other research into worms. That work, however, sadly tailed off once more 'exciting' science, subsidised by the war industry, kicked off after World War I.

Darwin was proved wrong on rocks sinking into the aerated soil, but right about the fact that a small amount of work, multiplied by a lot of workers, equals a lot of labour. How much soil do worms move around in a year? Darwin settled on 10,516 kilograms of 'dry earth' passing through worms per acre of land per year.[9] That's just under 26 tonnes of matter per hectare. An awful lot, in other words.

Epigeic earthworms, those that live in the leaf matter and the top of the soil, can ingest 3–50 milligrams (measured as dry matter) of dung, or any other kind of litter, per gram of earthworm per day.[10]

There's another group of worms, called geophagous worms ('earth-eating' worms), which can consume 200–6700 milligrams of soil per gram of earthworm per day.[11] This is a huge amount for a small creature; up to seven times their own body weight. But, it's still a tiny amount, about two teaspoons of soil. Multiply it out, though, and it adds up.

A lot of small creatures can do a lot of work, as Darwin noted. The biggest geophagous worms could, using the number of worms per hectare that Darwin estimated, chomp their way through 325 tonnes of soil every year. Even a field full of little epigeic worms could get through 2.4 tonnes in a year. Yes, these are best-case scenarios, but hey, worms!

What, exactly, is doing the work? Is it one species, or many? Turns out there are up to 6000 species of earthworms in the world, with about 1000 native to Australia,[12] and they all have similar functions to those Darwin observed in Britain. They spend their lives consuming dead matter. They eat fallen leaves. They consume dead grass and roots. They'll get into poo and break it down. (In fact, worms do much better in the presence of animal poo than otherwise, Italian researchers have found.)

Of all the indigenous worms around the globe, it's the Lumbricidae family, hailing from Britain, Europe and north Asia, that are most talked about. They're generally the ones we recognise in modern gardens, because they breed fast, can move quickly, and colonise areas successfully, given the right conditions. Most regions only have a handful of species that help in agriculture. It was a Lumbricidae member that gladdened Graeme Stevenson's heart on my farm. *A. longa* works in tandem with dung beetles, especially when the beetles hibernate in winter.

Despite Darwin's book, it wasn't until 2014 that the true role of worms was properly quantified. In a metadata analysis (a deep data summary of previous studies), Jan Willem van Groenigen led a research group that showed worms can increase crop yields by an astonishing 25 per cent over the long term.[13]

We now know a lot more about worms than even two decades ago – including that Darwin and Hensen's estimates of 133,000 earthworms per hectare of healthy soil fell way short. Tasmanian and

New Zealand research found that the most productive pastures in their worm trials had up to 7 million worms per hectare – worms weighing nearly two and a half tonnes in total. They also found that earthworms introduced to worm-free perennial pastures produced an initial increase of 70–80 per cent in pasture growth.[14]

How do worms do this miracle increase in yields? Humic acids, formed in the gut of worms, act almost like hormones, which enhance plant nutrition and growth. Worm-produced humic acids allow a plant's roots to grow longer, and put out more branches. New growth from more branches means greater ability to absorb nutrients. How this happens, in a chemical sense, is still a bit of a mystery.

Since about 1985, we've been pretty sure that humic acid can release phosphorus from soil and make it bioavailable – that is, available in a form that a plant can take up. And we now know that humic acid from worm guts is also really good at dissolving rock. So, worms don't just help plants to grow, they actively help make soil structure from rock crystals. And it probably happens faster than glaciation, in the right circumstances.

Worms also do a thing called bioturbation. The term bioturbation, meaning the action of animals to mix up soil and spread nutrients, was only coined in 1952,[15] seven decades after Darwin's book came out – and its technical definition was still up for debate in 2012.[16]

In essence, worms act as the engineers of soil structure. They're big enough to move matter around, and they move faster than fungi, plant roots and most other subterranean life forces. They can incorporate leaf litter into the soil, and through their actions of forming castes, burrowing, making middens and other activities, earthworms can significantly alter the physical properties of soil, as well as nutrient availability. At the same time, they alter its biological communities, affecting not only plant parasitic nematodes, but also above-ground plant communities. In short, they provide for more air in the soil, and better water penetration. They speed up root growth and improve plants' root structure, all the while moving significant amounts of minerals and organic matter, mixed through the worm's guts, throughout the structure of the soil.

Alongside Ceres, Demeter and Pachamama, ancient goddesses of the soil, perhaps we also need a god or goddess of worms.

Worms have another capacity. One group of researchers likens them to Prince Charming. That's because there are soil microbes known as Sleeping Beauties, which sit dormant, idly waiting for the right conditions to kick into life, waiting for their food to arrive – food in the form of dead organic matter, or a plant's roots. Dormant microbes can't move to where the food is, so they sit back, waiting for things to improve. Scientists wondered why evolution had created this seemingly flawed and paradoxical survival strategy – creating life that just sat there in the hope that food would come within reach. But then it was discovered that the right conditions just might include a 'kiss', according to a team led by George Brown, from the Mexican Department of Soil Biology, at the Institute of Ecology.[17] This is the 'kiss' of a worm.

Now, we've all had a kiss that might have been a bit wet, or sticky, and a worm's kiss sounds a bit like that. A worm's 'kiss' is to digest the microbe, or perhaps rub up against it, and secrete mucus onto it, in the form of a glycoprotein. What the bacteria get is a moist, mobile environment, some sugar to kickstart activity, and a free ride through the soil to more food. This wakes up the dormant microbes (which, in turn, help the worm to digest organic matter) – which causes a spurt of microbial activity that scientists have called a 'hot spot' in the soil.

Hot spots occur when a whole load of sleeping beauties wake up to a glycoprotein feast that they've been moved to, thanks to the worm. And these hot spots are a huge boon in terms of soil fertility.

In hindsight, perhaps I shouldn't have fried and eaten an earthworm. Perhaps I should've just let it kiss me.

Worms, those tiny but massively important underground engineers, whose mere presence implies healthy soil, are abundant in the right conditions. But how many did those researchers find on Tasmanian dairy farms? Seven million per hectare?[18] That has to be a misprint, no?

To relate that to the home gardener, that's about 700 squiggly, wriggly soil builders per square metre of your garden beds. The tonnes of organic matter they turn over in a year, aerating the soil as they go, is no mean feat, considering the effort it'd take a human, or the diesel you'd burn if you had to do it by machine – both of which would destroy soil life in the process.

Worms are the ultimate visible soil builders, capable of dissolving rock, feeding plants, digesting dead matter and aerating soil.

And, yes, there are lots, and lots, and lots of them in healthy soil.

Imagine my surprise, then, when I discovered the humble nematode, only about eight years ago, when we had our first soil tests on our farm. Worm-like creatures, but not actual worms, they represent – incredibly – 80 per cent of all multicellular animals on Earth.[19]

Most nematodes are invisible. They exist in and on us. They exist in plants and most animals – you're bound to have some parasitising you right now, in your gut and on your skin (are you wriggling around yet?). Nematodes are in and on just about everything, from the highest mountain to the world's deepest mine, 3.6 kilometres (over 2 miles) beneath the Earth's surface.

Nematodes represent 90 per cent of life on the ocean floor.[20] They have been found living in vinegar, and underneath beer taps in pubs. They have lived for 32 years in the fridge on wheat seeds.[21] That's not where they want to live, however. Relatively warm, moist, food-rich soil is a dream for them – which is why 90 per cent of all the world's nematodes live in the top 15 centimetres of soil.[22]

And if you linked all the nematodes on Earth from end to end, you'd not only be able to get to the Moon and back, you'd get past the sun. You'd line them up and go past the edge of our solar system.

You'd have a conga-line of nematodes all the way to our nearest star after the sun, Alpha Centauri.

You'd get there, over four light years away, with a line-up of all the nematodes currently living on Earth. There *and* back again![23]

That's how many nematodes there are, and how small they are. For every human on Earth, there are 60 billion nematodes. Wild, heh?[24]

Nematodes are pretty amazing, before we even look at their role in soil. A nematode endured the re-entry fireball when the Space Shuttle Columbia burned up shortly after take-off in 2003 – believed to be the first known life form to survive a virtually unprotected atmospheric descent to the Earth's surface.

And along with tardigrades, nematodes are likely to be the first multicellular life from Earth to visit our closest star system. Chosen by NASA, the nematode *Caenorhabditis elegans* could well be the first interstellar species to travel to (purely by coincidence) Alpha Centauri, through the University of California's Starlight program.

While some people may know of the few species of harmful nematodes that affect animal and plant health, that's only a tiny fraction of what's out there. Scientists estimate there are probably about a million species of nematodes[25] – though only a fraction are yet identified.

As N. A. Cobb, the father of nematology, who founded the scientific niche and identified 1000 species, wrote before he died in 1932:

> If all the matter in the universe except the nematodes were swept away, our world would still be dimly recognisable, and if, as disembodied spirits, we could then investigate it, we should find its mountains, hills, vales, rivers, lakes and oceans represented by a thin film of nematodes. The location of towns would be decipherable, since for every massing of human beings there would be a corresponding massing of certain nematodes. Trees would still stand in ghostly rows representing our streets and highways. The location of the various plants and animals would still be decipherable, and, had we sufficient knowledge, in many cases even their species could be determined by an examination of their erstwhile nematode parasites.[26]

Take away everything else, and nematodes would still form the outline of Earth and all on it, there are so many of them.

For every billion bacteria in healthy soil, there are only about 50 to 100 nematodes,[27] but they are much bigger in size than bacteria, and often make their meals out of bacteria. Each one can chomp through 5000 bacteria a minute on a good day.[28] That's right, crazy numbers.

Like all animals, nematodes feed off other things that are either already living, or have previously lived. Some eat poo. Some eat dead organic matter. A few thousand species are plant parasites. And while some nematodes can cause problems if they eat our crops – which are the ones we've often focused on in agriculture – most play an incredibly vital role.

Nematodes have four major functions in soil. The first is that they eat lots of bacteria and fungi, and so help regulate the populations of other soil organisms. The second is that they mineralise nutrients into plant-available forms; as we saw with fungi and bacteria, and just now with worms, they are plant feeders. Third, because they're not the biggest creature in the soil food web, they are consumed by other things – including bacteria and fungi! – and are another vital link in the food chain. Their fourth major role is that they consume disease-causing organisms.

In other words, on the whole, they do way more good than bad.

So, as we have seen, worms make soil. Microscopic worms – nematodes – far outnumber all other multicellular animals, and act as soil moderators, soil conditioners.

From the unseen, to the seen and disparaged, worms also make us. They form part of our individual ecosystem, and the system that feeds us.

Perhaps it should be left to Darwin, whose fame helped drag the worm from obscurity, to end this ode to the small, wriggly and dirty. Especially as he finishes his book with a more excited tone than he starts:

When we behold a wide, turf-covered expanse, we should remember that its smoothness, on which so much of its beauty depends, is mainly due to all the inequalities having been slowly levelled by worms ... It may be doubted whether there are many other animals which have played so important a part in the history of the world, as have these lowly organized creatures.[29]

CHAPTER 10

Bombs, Germs and Plants: 100 Years of Fast Fixes Creating Big Problems

Timing, as we all know, is everything. Whether you're a reality television star trying to become president, a café opening in the local high street, or a shepherd tending a flock. What you do matters, but timing can matter more.

This is a story of blockades, wars, rebuilding economies, and the belief in humans as the architects of nature. Fundamentally though, once again, it is a story about our inextricable dependence on soil.

But first, let's go back a bit.

When humans started farming, we paid more attention to soil than we do today. We could see the results, and feel it in our bellies, when soil was lacking, as well as when it was doing well. Civilisations have had a nasty habit of turning arable land into desert when they became too big and lost their ability to nurture soil; the Fertile Crescent is fertile no more. In those days, however, we always had more earth. We had more land we could move to.

When Darwin's worm book came out in 1881, British gardening was at a relative high point. In the Victorian era, gardeners were growing pineapples in English greenhouses. Exotic fruits and vegetables had been arriving from warmer and more distant climes for a few centuries. A good gardener knew that worms meant lively soil, even if they couldn't see the tens of billions of microscopic workers that also formed their soil ecosystem.

Growing food was at once both a noble act and peasant labour. Those peasants, those home gardeners (many partaking out of necessity), and those landed gentry of Darwin's ilk who tended plants and paddocks, they could see fertility. They could smell it. At times the mechanisms eluded them, but they all desired healthier soil because they knew healthy soil grew bigger, better-tasting vegetables.

Healthy soil meant an easier, more profitable life.

By that stage scientists, including Justus von Liebig, had worked out that nitrogen is a key macronutrient for plants, making them grow bigger, faster. Indeed, few things have such immediate and visible impact on plant growth as nitrogen. They also knew that soil nitrogen could be supplemented with nitrogen-rich animal poo and urea (found in urine). They realised, however, that concentrated natural nitrogen reserves, such as potassium nitrate (or saltpetre, as it was known), were in small and fast-dwindling supply. The amount of manure was also limited, as well as being hard to transport in the era before cars and trucks.

There's an essentially unlimited source of nitrogen in the air (the atmosphere is nearly 80 per cent nitrogen), but it's hard to capture in usable form. Scientists were keen to figure out how to harness atmospheric nitrogen and turn it into reactive nitrogen, both because it could be used as an artificial fertiliser, and because reactive nitrogen (as opposed to stable atmospheric nitrogen) is great for making bombs. Indeed, reactive nitrogen can be a powerful explosive; some of the original saltpetre that was mined was used for gunpowder.

Unfortunately for many, in 1909, a German gentleman named Fritz Haber worked out that an input of mass amounts of energy, at high temperature and pressure, along with a catalyst, could trap atmospheric nitrogen and turn it into ammonia.

It only took a year for Carl Bosch, from the company BASF, to discover a way to ramp up production for commercial use, in both bombs and farms, and by 1913, the first ammonia factory was up and running.

Then came World War I. And with it came the desire for bigger, better, more deadly bombs.

Timing, indeed, was everything. The rest, as they say, is history.

The German war machine geared up to turn atmospheric nitrogen into bombs. Factories all over Germany – especially following the British naval blockade of Chilean saltpetre (a natural form of nitrate) – turned their attention to making ammonia. It was very much to Germany's advantage that the Haber–Bosch process, as it became known, was available during the war, even if it didn't ensure victory. No longer did they have to rely on saltpetre for gunpowder, and animal waste for fertiliser to feed the war effort. Ammonia could provide both.

(Incidentally, Fritz Haber, a strident nationalist, spent the war years devising chemical weapons – including chlorine gas, which, when unleashed on the French in Ypres, killed at least 6000 men in a few minutes, and injured 10,000 more.[1] He later 'improved' on it by inventing mustard gas. His wife, also a chemist, committed suicide in the family home a week after his first poison gas strike in 1915, which she considered 'immoral'.)

Haber was awarded the Nobel Prize in Chemistry in 1918, though not without controversy, considering he was known by the Allies as The Father of Chemical Warfare. (This was a bit hypocritical, considering the Allies ended up using more chemical weapons before the war's end than the Germans, even if the Germans did use them first. But there you go.)

After the war, Germany was a broken, impoverished nation that had spent all its efforts attempting to conquer its neighbours, and had to rebuild. Those ammonia factories sat waiting and ready to make hundreds of tonnes of readily available nitrogen. Soon, they became the foundation of the fledgling fertiliser industry. 'Bread from Air' was a slogan the Haber Institute, an arm of BASF, used to promote Fritz Haber's work on artificial fertilisers.

(As a second side note, Haber continued work on poison gases after the First World War. He invented Zyklon B in the 1920s, which was eventually used in Hitler's gas chambers. Haber, a Jew himself, died in exile in Switzerland, but it is estimated his brainchild Zyklon B killed over a million of his compatriots.[2])

After the Second World War, factories all over the globe were repurposed. Nitrogen was now readily available as fertiliser instead of bombs. Tank factories were re-tooled to make tractors.

Now, war stories aside – and despite the fact that about 100 million people have died globally from guns and bombs armed using ammonia from a process that Haber invented[3] – it's Haber's fertiliser that has had a bigger impact. Billions of people probably owe their lives to it.

Why? Because of plants' love affair with nitrogen. In fact, today about one mouthful out of every two a human eats is a result of the nitrogen used in artificial fertilisers, pretty much all of it made using the Haber–Bosch process.

Nitrogen is like a growth promotant for plants. It's vital to form plant structure, and its synthesis spawned what came next: the Green Revolution. No longer at the mercy of natural soil composition, or of crop rotations, the Haber–Bosch process allowed a flourishing industry on farmland. We've been able to expand agriculture into areas previously considered marginal. We've been able to increase yields in lots of crops, especially grain, while also growing more grass for dairy cows. Industrial-scale agriculture couldn't exist without ammonia. Haber's ability to trap and harness atmospheric nitrogen has fuelled massive growth in monoculture crops, which has enabled feedlots, fish farms, 60,000-sow piggeries, battery hens, and – many would argue – a corresponding rise in the human population.

The problem, of course, is that nothing comes for nothing. The problem isn't with Haber's ammonia, but more with how it's made (using extraordinary amounts of fossil fuels), and what it does to the environment.

In short, all those beautiful soil bugs that we met in earlier chapters tend to get turned off by nitrogen in artificial fertiliser. A big fat injection of Haber's nitrogen, if it doesn't kill the microbes, certainly makes for a lazy subterranean ecosystem. And just like a sugar rush for a human, there's a post-nitrogen slump. The quickest, easiest way to get over the slump is to add more nitrogen. But it's a losing game.

And with all the gains in plant growth, there's been a corresponding decrease in the nutritional quality of our food – and at an escalating environmental cost.

While there's no lack of nitrogen to be harnessed from the air, other elements identified early on as vital to plant health are less easy to find. The second most vital macronutrient for plants is phosphorus. Phosphorus is super effective at enhancing shoot and root growth, amongst other functions such as aiding cell division and efficient energy use.

Phosphorus exists pretty much everywhere in rocks, but some extraordinarily rich deposits around the globe have proven a boon to industrial agriculture, because the more nitrogen you put on your soil, the more phosphorus it's good to have available, too. Each helps the uptake of the other.

One such phosphorus deposit acts as a cautionary tale of what can go wrong if we don't treat the Earth right.

There's a speck on the global map that was once known for its phosphorus. It's a dot, 21 kilometres (8 miles) square in size. An island that lies nearly 9000 kilometres (5500 miles) south-west of Los Angeles, more than 4000 kilometres (2500 miles) north-east of Sydney, and 300 kilometres (180 miles) from its nearest neighbour, Kiribati. That speck is Nauru, a tiny Pacific island of 10,000 people that is fringed with palm trees, but whose heart is now quite literally dead.

Nauru was once a tropical paradise. Before that, it was a toilet.

For an estimated 4 million years, prior to human habitation, Nauru was where a lot of seagulls emptied their guts.[4] Over millennia, the birds' poo accumulated, then calcified. After being inhabited by

Polynesians and Melanesians for an estimated 3000 years, a European prospector, Albert Fuller Ellis, arrived in 1900. Ellis found what looked like calcified wood on the island, but turned out to be ancient seagull poo – and the richest source of phosphorus ever discovered.

A persuasive alliance of three Commonwealth nations, Australia, New Zealand and Britain, stripped 43 million tonnes of phosphate-rich earth from Nauru[5] and shipped it off to be used in garden centres, orchards and fertiliser factories the world over. Then, of course, the companies moved on.

Today, Nauru's inland is a wasteland of stalagmite-like peaks, washed of nutrients by tropical rains. To this day the land remains poisoned with the heavy metal cadmium, after subsoils and cadmium-rich rocks were exposed to the air. Phosphorus in concentrated form is often found in rocks that also contain cadmium, all around the globe.

It's a sad irony that a place that only a short time ago boasted one of the most sought-after ingredients for fertiliser, to grow better crops, has land so degraded, and health outcomes so poor – partly because they can't grow food themselves. Today, Nauruans have a life expectancy about 20 years shorter than most developed nations, along with astonishingly high rates of heart disease, diabetes and kidney failure.[6] Their only real source of income these days is as an off-shore processing facility for asylum seekers attempting to reach Australia by boat.

Of course, this story is just one of many when you look at what we do to 'improve' our soil. Science finds the magic bullet, and we want to get as much of it as possible into our gardens or paddocks.

Be it nitrogen, phosphorus or potassium, or be it seaweed, rock dust or selenium, we will eventually run out of everything that isn't already on Earth, unless it can rebuild or regrow. Concentrated sources of some minerals allow us to use them quicker. But the real, long-term fix is in letting soil heal itself. If it isn't in a natural cycle, we're just buying time.

The thing is, phosphorus is already present in most agricultural soils. It is, after all, present in most rocks on Earth. But often it's locked up, in a form that plants can't readily access. Many of the world's soils, even old depleted soils, contain enough phosphorus for centuries of farming without needing to add more. We just have to come up with systems for releasing it in plant-available form.

Soil life can do just that.

For the last hundred years we've focused on ways to extract the main macronutrients for plants, and lauded human ingenuity in the way we obtain and utilise them. We've championed NPK because the synthesis of nitrogen, the finding of rich deposits of phosphorus, and the widespread acceptance of artificial fertiliser have formed the basis of the Green Revolution. But, as we'll see in the next chapter, their use has actually helped turn the world brown.

And in the process, it has increased – rather than decreased – nutrient deficiency on a frightening global scale.

CHAPTER 11

How the Green Revolution is Turning the World Brown

Little Red Hen. That's what I felt like. On a sunny but bitingly cold August day in 2018, I rolled up a tarp that was suppressing weeds on the western edge of our market garden and planted wheat. Nadia, our gardener who hails from Western Australia, home of the nation's biggest wheat belt, didn't help. *Wouldn't* help. She had no interest in preparing, by hand, 60 metres of garden bed to house wheat. My partner Sadie told me I was on my own in hand-sowing the French ruby variety. Our cooks scoffed when asked if they'd help hand-grind it. Just like the storybook character Little Red Hen, whose barnyard companions didn't want to do the heavy lifting, everybody found an excuse to avoid sowing, weeding, harvesting, threshing, winnowing and grinding our wheat.

I baked the loaf alone, too. But no sooner had it come from the oven than everybody wanted a piece. What does really fresh, spray-free wheat taste like in bread? Bloody wonderful, that's what. A warm, nutty flavour, with aromas reminiscent of antique wood. Caramel notes. Deep, primordial hints, which satisfied on more than one level. Yes, I remember the flavour. But I also remember the hours, about 120 in all, that went into growing and preparing grain – enough for about ten loaves – for the table.

That moment, when you realise that growing wheat and turning it into flour, the usually cheap stuff you take for granted, is an enormous drain on your time and energy, is telling.

I'm not sure I'd ever do it again, but it did give me huge respect for what humans got themselves into when they stopped roaming and started farming.

Now, I have to admit, it's not the Dark Ages. The seeds I planted were modern wheat, able to hold 40 seeds per plant head, compared to the 4–6 seeds wheat boasted back then. We have strong metal tools with long handles to aid our backs. We have a well-engineered hand mill. And still the labour was extraordinary. And the amount of wheat we eventually got from the crop was able to sustain a family for only two weeks.

Anyone who has grown wheat by hand knows that farming, on a subsistence level, is hard, unrelenting yakka. Growing food, for many, is a lifelong task.

Very few of us want to be Little Red Hen. Countless times over the years, we humans have innovated to improve our lot. We learned that the root of Queen Anne's lace, a plant originally kept for leaves and seeds, could be bred to become fatter and sweeter – and so was born the modern carrot. We discovered that some barley stalks grew way more seeds in their heads, so we selected for those. We bred for bigger, more flavoursome, sturdier fruits and vegetables. We did all of that, saving the best seeds, learning from our mistakes. Farmers suffered the vagaries of nature, hoping – and usually praying – that the harvest would be good.

Fast-forward to the 21st century, and by comparison, for many farmers, the harvests are relatively good. Mostly we can thank not just Fritz Haber and his ability to draw nitrogen out of the air, but also an Iowan-born American, Norman Borlaug, who took plant breeding to new heights.

Borlaug's name isn't on everyone's lips. And yet, this trained forester turned agronomist has had arguably the largest impact on agricultural land, on soil, and on our diet than anybody else in human history. As the person who first popularised the term 'Green Revolution' for the technological advancements of agriculture from the 1950s, Borlaug was a man who put his science where his mouth was.

Borlaug was no Haber, no maker of chemical weapons. But like Haber, Borlaug believed in the ingenuity of the human mind, and our ability to find solutions to agricultural problems. And like Haber, he was enamoured of the ability of ammonia, artificial nitrogen fertiliser, to get plants to grow really quickly. Working first with wheat, then with other crops, Borlaug wanted to find better ways to feed the world.

Through plant breeding, selecting for new types of crops and ones that could be grown in the new high-nitrogen environment aided by the Haber–Bosch process, Borlaug set about transforming agriculture. He spent all his working life, really, helping poorer countries select higher-yielding crops, and plants more resistant to disease.

Wheat, one of the three most commonly grown crops on the planet, is a great example.

Traditionally, wheat was much taller than most of it is today.[1] But when you add artificial fertiliser to it – nitrogen from the Haber–Bosch process – the plant gets too heavy for its stalk, and some stalks bend over. The plant also puts more energy from photosynthesis into growing the stalk (which we can't eat, as it's indigestible to humans), rather than the head.

The logical thing would be to grow shorter (dwarfing) varieties, so the plant can put more effort into the grain, and less into the stalk, and there's less chance of the stalk breaking and the seed being wasted.

So far, so good. Borlaug's work on plant breeding, especially dwarfing varieties, was super successful. It increased yields, reduced disease, and allowed grain to be grown on impoverished, nitrogen-poor soil.

No longer did farmers have to plant nitrogen-fixing crops in alternate years; they could now grow wheat every year. By some accounts, Mexico's wheat harvest tripled in the decades after the new technologies came into effect.[2]

It was so successful that the same principle was applied to rice. After increasing wheat yields in India and Pakistan, the Green Revolution's effect on rice production was dramatic, if not quite so large. According to recent evidence, the productivity gains from seed selection were about 1 per cent per annum for wheat, and about 0.8 per cent for rice across the board.[3]

Using dwarfing wheat, you can reduce the size of the plant by 20 per cent, and increase the yield (the actual, usable wheat seed) by 5–10 per cent.[4] A shorter plant is, in short, way more efficient at producing grain.

But – and this is the big but – the protein content of the wheat, compared to a virtually identical non-dwarfing wheat, can be 12 per cent lower.[5]

You grow more, but get less in some ways. More calories, less protein.

Why does this happen? Change the conditions in soil, and you'll change the food that the soil grows.

Remember the rhizosphere we met earlier? That's the magic region around a plant's roots where most of the bacterial and fungal action occurs. It's the place where a plant trades sugars and amino acids, and up to 100,000 different compounds, in exchange for nutrients. It's where most of the goodness for a plant is sourced.

Well, when you breed for one characteristic in a plant, you also breed for other, unknown factors. And dwarfing wheat, according to research that came out in 2020, has a less complex, smaller, weaker rhizosphere.

In a paper in *Nature*, a team of researchers led by Vanessa Kavamura[6] showed that you actually change the ecosystem around the roots because of the plant variety. It's a complex report. But, to put it simply, dwarfing wheat can't sustain the same vibrant, complex underground ecosystem that taller wheat can. There are fewer of those wonderful soil microbes, communicating less and doing less of what they are designed to do – to truly nourish and protect plants. Choosing for dwarfing wheat has dumbed down, and limited, soil life.

That's Green Revolution breeding. But what about the additional effect of inorganic nitrogen, via Haber–Bosch, which those new varieties were also bred for? Well, two years prior to her research on wheat's rhizosphere, Kavamura looked into just that.[7] And she found that if you add nitrogen in the form of artificial fertiliser, you'll also decrease microbial activity. Nitrogen is like kryptonite for soil bugs. Add nitrogen and you'll switch off, weaken or even kill soil life. Change breeds, and you can alter soil life.

A long line of research shows similar results. The very things that Borlaug was doing to increase the amount of food grown – hybrids, plant size and artificial fertilisers – have damaged soil.

More than that, research shows that photosynthesis of dwarfing wheat can be 18 per cent lower.[8] Remember the underground economy? Borlaug's efforts inadvertently cut down the amount of photosynthesis the wheat could do, which in turn reduces the sugars available to soil microbes, and hence decreases soil carbon. Less photosynthesis means less carbon in the soil, less glomalin, fewer microbes, and poorer soil health. Despite calling it the 'Green Revolution', much of Borlaug's work actually created less greenery on crops, and less of the beneficial sugars that greenery produces.[9] It cuts off the food supply to soil at the source.

It's not just developing nations that were the beneficiary of this new technology that inadvertently undermined soil health. The United States, Australia and Europe all chased the dream of more efficient crops that didn't rely on soil health or soil carbon. It's no surprise that Borlaug's native Iowa, as you may recall from Chapter 8, *Here Today, Gone Tomorrow*, lost half its topsoil in the 1900s, during his watch.[10] In large part that was as a result of artificial fertiliser and crop selection on American agricultural land.

Through the Green Revolution, humans have created simpler, less stable ecosystems around the roots of dwarfing plants, and turned off all the bacterial richness – the soil's ecosystem – in the process.

As we saw above, the Green Revolution has also reduced the nutrient density of our food. By ignoring soil, and aiming for yield, it has unwittingly decreased a macronutrient of wheat, protein – which may not matter much to many of us in richer nations, who are over-nourished anyway, but matters a lot more to those on the margins, who are borderline nutrient deficient.

Now, this drastic drop in wheat protein is just one macronutrient. Imagine if we could test for all the micronutrients that the Green Revolution has diluted in food. The phytochemicals, the antioxidants, the aromatic esters.

Borlaug's aim was to feed the world. And in terms of calories, he's done a pretty good job. It's estimated that over a billion people are alive today because of the work he's done to increase the productivity of agricultural land.[11] But it has come at a cost in terms of nutrient density. Not only can we now grow food in soil lacking in the natural ability to provide nutrients, but the crops themselves are limited in their nutrient profile.

The long-term issues for soil are correspondingly huge. Partly because the Green Revolution and artificial nitrogen are inextricably entwined, Borlaug's work has enabled the planting of crop mono-cultures, the spraying of herbicides and encouraged unfettered use of the plough. More tilling of soil, leading to more erosion.

Because it has facilitated the wholesale use of reactive nitrogen, which has reduced soil life, it hasn't been good for the health of soil, the planet, or us.

When you spread artificial nitrogen on a field or a paddock, about 60 per cent of it is never used by a plant; some say closer to 70 per cent.[12] That's 60 per cent of a farmer's money spent on something that does their crop no good.

Nitrogen in the air is stable, it's inert. When turned into fertiliser it becomes reactive, meaning it can be changed, and reacts more with other things, like oxygen to produce nitrous oxide, or within a plant to produce surplus nitrates.

It's not just the target plants that it reacts with, of course. Nitrogen is leached into waterways, where it becomes a toxin in any quantity. Over-nutrified water becomes de-oxygenated when rampant algal growth sucks oxygen from the water, leading to dead zones.

Around the world, vast bodies of water are constantly being poisoned by fertiliser use. Between 1990 and 2004, it was estimated that the area of ocean adversely affected by nitrogen doubled.[13] In fact, according to a report in *Science* magazine, which covered up until about 2008, 'Dead zones in the coastal oceans have spread exponentially since the 1960s.'[14] EXPONENTIALLY! In the Gulf of Mexico, 41 per cent of the nitrogen causing barren patches is from fertilisers.[15] In 2017, an area the size of New Jersey was devoid of aquatic life in the same gulf.[16] As American soil scientist Dr Kris Nichols puts it, 'The last thing the gulf needs is another dead zone.'[17]

Globally, these 'dead zones' have now been reported from more than 400 waterway systems, affecting a total area of more than 245,000 square kilometres.[18] That's an area bigger than the United Kingdom.

The other effect of artificial nitrogen, which is largely unheralded outside of scientific circles, is its role in the atmosphere: it becomes nitrous oxide. Not all of that 70 per cent of the nitrogen added to soil to grow crops is washed away. Some is blown away.

Nitrous oxide, N_2O, otherwise known as laughing gas, is no laughing matter. It's a very potent greenhouse gas, with a way longer half-life than methane. Depending on where you get your data, it's estimated to be about 265 to 310 times more warming than carbon dioxide,[19] and persists in the atmosphere for centuries (unlike methane's decade or so).

Nitrous oxide is always emitted when you use artificial fertiliser. What does this do? Well, according to a group of scientists from the universities of Melbourne, Virginia and New Hampshire:

It escapes into the environment – it cascades through atmospheric, terrestrial, aquatic and marine pools. In the atmosphere, reactive nitrogen leads to smog, acid rain, intensifying the greenhouse effect and stratospheric ozone depletion. In terrestrial ecosystems, reactive nitrogen leads to soil acidification, forest dieback, and biodiversity loss.

In marine and freshwater ecosystems, reactive nitrogen contributes to freshwater acidification, groundwater pollution, ocean acidification and eutrophication – a build-up of nutrients in water that causes dense algae growth.

High levels of reactive nitrogen in water and air have been directly and indirectly connected with human diseases and allergies like increased incidence of asthma and colon cancer.[20]

Now, to be fair, nitrous oxide – reactive nitrogen – can come from other sources besides artificial fertiliser. Most – some say about 62 per cent – of all nitrous oxide emissions come from natural sources: soil (the biggest emitter), oceans, forests.[21] We all release a bit from our bodies. There's reactive nitrogen in poo, in dung. A bit comes from plants.

But it's important to look at why, only now, we're getting aquatic dead zones. And why atmospheric nitrous oxide levels have increased about 20 per cent since the Industrial Revolution.[22] Thanks to human activities, there's been at least a doubling of human-induced reactive nitrogen in the world in the last couple of hundred years[23] – and the lion's share of this rise in nitrous oxide comes from artificial fertiliser, intensive animal husbandry and fossil fuels.

Atmospheric nitrous oxide is at the highest level we've seen for 800,000 years – about four times as long as there have been modern humans.[24]

'But Matthew, you seem to be overly critical of the Green Revolution,' I hear you say. 'It fed people, and that isn't bad. We need nitrogen to guarantee fertile fields and viable crops, don't we?'

Well, let's take a look at the numbers.

Thanks to Borlaug, the production of cereal crops tripled during the fifty years to 2012, with only a 30 per cent increase in land area cultivated.[25] According to a review of the literature in 2013, Green Revolution research and implementation saved an estimated 18–27 million hectares from being brought into agricultural production.[26] That's forests and grasslands that weren't ploughed to grow crops.

A 2012 review found:

> Between 1960 and 1990, food supply in developing countries increased 12–13%. Estimates suggest that, without … crop germplasm (ie seed variety) improvement efforts, food production in developing countries would have been almost 20% lower (requiring another 20–25 million hectares of land under cultivation worldwide). World food and feed prices would have been 35–65% higher, and average caloric availability would have declined by 11–13%.[27]

However, a lot of those crops weren't used for humans to eat.

Yes, we feed a lot more people, thanks to the Green Revolution, and the grains we produce are cheaper. But – and this is important – what we have actually done is to subsidise fossil fuels. In the United States, 40 per cent of corn is grown specifically to be turned into ethanol,[28] with the by-product fed to intensively farmed animals. So, nearly half the land 'saved' from the Green Revolution by growing 'food' cheaply was put to turning the 'food' (corn) into ethanol, then using the waste as animal fodder.

In fact, it's worse than this, because many grains are grown solely to feed to livestock – about 30 million hectares in the United States alone are grown specifically to be fed to domesticated animals.[29]

It's not just corn, of course. Soybeans are also used to make ethanol, and the meal used for livestock fodder. In fact, over half the total grain output of the United States, and over 40 per cent of the global grain production, is fed to livestock.[30]

Globally, we use nearly 30 million hectares of land specifically to grow ethanol crops[31] (with the waste fed to animals in intensive farms).

It's bad maths to simply say we fed the world using the Green Revolution. We ended up feeding ethanol plants and factory farms, not people. Ethanol production uses more land than was 'saved'.

In Borlaug's early career, the Green Revolution might have been all about feeding people, but today it's clear that its technology isn't saving farming land. It's ruining it. We simply grow staple crops at much higher rates, in a soil-ruining manner, doing it much cheaper, just so we can use them for things other than human food.

Borlaug has even had a low-protein, high-yielding wheat named after him to celebrate the 100th anniversary of his birth, Borlaug 100. As one grower of Borlaug 100 puts it, 'feedlots are our only market. They want as much wheat as they can get, and they don't care what the protein is.'[32]

I keep wondering, have we ever actually considered whether artificial nitrogen and one-dimensional crop selection were really the best use of things to hand? Or could we have done better, if we'd known better? Just because we've had measurable results doesn't mean we got it entirely right.

Hindsight's a great thing, but foresight can be more useful if you recognise past failings.

For a start, we've squandered some of the best fertiliser that was right under our noses. We're not using animal waste the way we could be. The use of fossil fuel–based fertilisers doesn't produce the results that animal poo can. Animals upcycle nutrients, and so their waste can be utilised better in the ideal farming system.

For instance, in 2017, a team led by Professor Deli Chen at the University of Melbourne analysed the data from 141 studies on how animal waste – including cattle, pig and poultry manure – could replace the use of synthetic nitrogen fertilisers.[33] They found that 'crop yields can be boosted by 12.7 per cent when between half and three-quarters of synthetic fertiliser is replaced with animal waste'. That's on top of the advances in seed selection. These results are above and beyond artificial fertiliser use, the supposed magical input for agriculture.

That same team also found that when manure is used instead of synthetic fertiliser, ammonia emissions fell by a very healthy 27 per cent; nitrogen leaching into groundwater dropped by a substantial 29 per cent; and nitrogen run-off also decreased by a none too shabby 26 per cent.

You increase yields, and reduce environmental damage, using natural fertiliser. That's using manure – but properly made compost has even greater long-term capacity to reduce run-off. Can you imagine what the benefit, long term, would have been to soil if we'd used natural fertiliser for the last 80 years instead of reactive nitrogen?

Imagine how happy Darwin's worms would be if we used natural fertiliser and poo, not nitrogen trapped using natural gas.

Could we have fed lots of people *and* improved soil if we'd thought of soil differently from the beginning of the twentieth century? Probably.

We've seen that every time you add inorganic nitrogen – in chemical terms, reactive nitrogen – to soil, you damage the world. Sometimes that may be worth it, to get some kind of dynamic boost to crops in impoverished soil, to start plants photosynthesising where they weren't before. But mostly, nitrogen harnessed using the Haber–Bosch process has lined the pockets of fertiliser manufacturers, while depleting the world's most precious growing medium – all the while damaging the air we breathe, the water we drink.

On top of this, it's estimated that 3 per cent of our carbon emissions come from making the energy-intensive fertiliser in the first place.[34] This isn't cyclical carbon, as in plants and animals, which is carbon that is already in play. This is fossil fuel carbon from natural gas – carbon that has been stored away underground for millions of years from a time when the atmosphere wasn't conducive to the lives *Homo sapiens* and other mammals enjoy today.

Of course, plants require more than just nitrogen. They also do well with liberal doses of phosphorus and potassium.

So what about phosphorus? Is it as bad for the environment as nitrogen? We know how bad it was for Nauru, but what about the places the phosphorus ended up?

Well, adding phosphorus to farming land isn't as bad as mining it is. But, it turns out, too much phosphorus is also a bad thing. While nitrogen is quick to disperse, phosphorus takes longer to leach into the environment – but when it does, it causes terrible damage, particularly to waterways. Unlike nitrogen, which tends to turn coastal waters into dead zones, phosphorus does it further upstream, over-nutrifying (and essentially killing) freshwater bodies, lakes and creeks and streams and rivers.

Since the pre-industrial era, the amount of phosphorus transferred from soils to waterways every year has more than doubled.[35] In agriculture, grain production had the largest contribution to the phosphorus load (31 per cent), followed by fruits, vegetables and oil crops, each contributing 15 per cent.[36] In other words, growing food, using phosphorus, is bad for the environment.

This isn't just a problem from the 1960s and 1970s – the height of the Green Revolution – that we now have to live with. It's still happening today. From 2002 to 2010, the phosphorus load from agriculture grew by a massive 27 per cent worldwide.[37]

A 2017 study seems to suggest that the amount of land that is taking in more phosphorus than it loses covers about 38 per cent of the global land surface, 37 per cent of global river discharge, and is home to about 90 per cent of the global population.[38] In other words, we're poisoning the land and waterways where 90 per cent of us live. Humans are putting an amount four times greater than the weight of New York's Empire State Building into the world's major freshwater basins every year.[39] You can't do that kind of thing for long before nature fights back.

Research in New Zealand points to another problem with phosphorus added to soil. It turns out the same cadmium that has toxified Nauruan soils has accumulated in the dairy paddocks of New Zealand,[40] thanks to concentrated deposits of rock phosphate virtually always being bound up with cadmium in the Earth's crust.

Globally, artificial fertiliser has at least doubled the amount of nitrogen and phosphorus escaping into the environment.[41] It's hard to quantify just how much damage its use has done to soil. So much of what we now know about soil wasn't factored into historical studies, and to model that on a global scale would be some undertaking.

What we do know is that Borlaug's Green Revolution did start out helping to feed a lot of people. It probably did prevent at least some land being turned over to agriculture to feed an ever-increasing global population, at least in the short term. But we also know there were probably better ways of doing things that we ignored while the fertiliser was cheap, and the rewards easy. It's something we'll be long paying the price for – not just in soil degradation and carbon emissions, but also in nutritional deficiencies.

If my wheat experiment taught me nothing else, it's that growing food the really old-fashioned way, even with modern seed varieties and tools, is phenomenally hard. The Green Revolution may have freed some of us from that drudgery and poverty, but it has also swung the pendulum in favour of large agricultural firms and their technology, taking the seeds and means to save soil out of the hands of the people who actually work the land.

Cheap grain has spurred on other industries, from intensive livestock, to ethanol. A whole swathe of the world's population lives with nutrient deficiency at a time of surplus food, and – oddly but not unexpectedly, when you consider the science – some of that deficiency includes the richer countries of the world.

All the while, soil has paid the price.

The Green Revolution has another, insidious impact. It's not just what we grow and what we add to the soil. It's how we tend it physically. As more and more land has been made available to grow crops, and is turned by the plough, something else has been going on.

CHAPTER 12

You'll Never Plough a Field by Turning it Over in Your Mind

Hal jar jami tar.
'Land belongs to the man who drives the plough.'
(18th century Indian slogan)

If there's one thing our forebears have revered almost as much as their gods of soil, it's the plough. Born of some ingenious human mind some 5500 years ago,[1] it lives in the imagination of the non-farmer in modern times – the sight of neat, ploughed fields, the furrows deep, the toil of the farmer obvious. It lives on in our language. We plough on, when times are tough. We plough through when we've got lots to do. When wars were won, the soldiers returned to the land, beating swords into ploughshares. The plough has a strong place in our collective psyche, the ultimate tool for growing food.

The Danish have a saying, 'To the priest his book, to the peasant his plough'. It is the plough we have to thank for easing many human backs, and filling many human tummies. It is the act of digging soil, turning it over, that has enabled much of what we think of as farming. Ploughing land loosened soil, and released nutrients more readily, especially nitrogen. It lessened weeds, gave furrows – grooves – for sowing seed, and was thought to allow more water infiltration.

After all, it's easier to imagine water penetrating loosened ploughed soil than the flattened ground of the untilled field. As some say, 'Plough deep whilst sluggards sleep, and you shall have corn to sell and to keep.'

And yet, and yet ...

It is the plough that has done arguably the most damage to our planet. Turning soil, it turns out, has done way more harm than any could have imagined.

It has unleashed at least a third of the increased carbon dioxide in the atmosphere that has appeared since the Industrial Revolution. The plough was the single biggest emitter of human-induced carbon into the atmosphere up until about the 1950s.[2] Bigger than coal, bigger than oil.

How does the plough do so much damage?

We've seen how soil is built, from the physical elements (crushed-up rock), to the biological (all those glorious microbes, worms, collembolans and more), and the chemical (how the biology of soil can make the nutrients bioavailable).

To understand what we've done to soil, it takes little more than an exploration of forests to see a record of decline. Virtually anywhere that agricultural land was once a forest, if there is still virgin forest next door, the difference is stark. Forests have stable carbon, and relatively stable soil life. Agricultural land doesn't. Arable land, the stuff we use for growing crops, is usually more carbon poor, and unstable by comparison. It's often bereft of the complexity of life, both above and below ground.

Visible differences can be seen on or over fences. In lines where tractors don't pass, the soil is more friable. It holds more air, and has more critters in it and on it. If it's pasture, there's often more variety there.

In his book *Call of the Reed Warbler*,[3] Australian farmer Charles Massy describes the thick, lush grasslands of Victoria that were stripped of vegetation, and the soil of fertility, only a handful of years after European-style farming arrived. In his exploration of Indigenous Australian agriculture, *Dark Emu*,[4] Bruce Pascoe recounts over and over, using settler diaries, how amazing the Australian landscape looked prior to colonisation. Horses sinking into friable soil up to their fetlocks, in areas now predominantly dust bowls. Both Massy and Pascoe call for a rethinking of how we manage agricultural land, including grazing animals.

Of course, the Australian landscape has changed really quickly, thanks to the relatively recent introduction of new plants, animals and land management practices. But farming has been, traditionally, really bad for soil in many places. Grow food and you mine soil, some say. By removing the plants, or the animals that have eaten the plants, you remove nutrients. Ploughing not only leaves soil vulnerable to erosion, it also kills the complex ecosystem underground.

We've known for aeons that ploughing land can cause damage. But just how much damage is now becoming clear. To understand the damage, we'll have to consider many things.

First is the physical. As we saw earlier, erosion is a massive problem. We're losing soil, on a global scale, on ploughed land, about 100 times faster than it can be made, according to the International Panel on Climate Change's 2019 Land Use Report.[5] Wind and water erosion take hold on the ground laid bare by the plough. There's also colluvial erosion, where loose soil moves downhill due to the action of gravity. Digging helps promote loss of the sand, silt and clay that make up over 95 per cent of soil. Once exposed, soil is too readily washed and blown away, or carried downhill.

But worse than this. Ploughing, we now know, sabotages something far more delicate – those things invisible to the naked eye.

Every time you plough a field, or dig over your garden, it's not just the visible soil that's in trouble, but the soil ecosystem. All those bacteria, protists, nematodes, archaea, algae and fungi, as well as the bigger life, the springtails, rotifers and worms, their home is ruined. Totally stuffed. It can be rebuilt, but it may never be the same. Just like a city, a rebuilt one may differ in lots of ways.

But a few things happen in the interim. Firstly, when you expose this microbial community to the air, lots of them die. When they die, they release carbon, which isn't replaced in the soil by other living microbes, at least in the short term. Cut open soil, and you release a whole bunch of stored carbon and nitrogen from the rotting vegetable matter as well. But – and this is the kicker – you also cut the hyphal threads, those long, invisible strands of fungal matter that provide all that underground connectivity in terms of nutrients and communication.

These hyphae are the source of glomalin, that magical superglue of soil. Slice through them, and soil structure is broken.

Now, glomalin was invisible to science until about 25 years ago, partly because it is so stable. Glomalin can last over 40 years,[6] but the slow depletion of the thing that stores most soil carbon, and does most of the heavy lifting in soil structure – well, that depletion starts in earnest the moment soil is dug. Glomalin's resilience has, perhaps, allowed us to damage soil with seeming impunity. A grower's lifetime is but a blip in terms of soil. It would be easy to blame other things – vermin, seasons, the gods – for 40 years of ever so slightly decreasing yields, rather than the plough. And that's what we did. The plough, after all, was revered.

For centuries a few cultures, independently of each other, recognised the most obvious damage from the plough. They found that rotating crops, planting different things in different years or seasons – in particular legumes to replenish nitrogen – helped. What also helped was a year or two of rest, usually letting the paddock go back to pasture, and allowing animals to graze it. Without realising it, they were letting the underground community regroup and repair.

If we'd worked that out centuries back, then why plough on? We like ploughing because it reduces the weed load, so the planted seeds don't get outcompeted. We like it because the initial digging gives us that hit of nitrogen for our crops. Ploughing has tamed formerly wild lands and made growing food more manageable, smoothing out bumpy ground.

Often, with your home garden, you'll still hear people suggest that you dig over the beds between harvests, to 'aerate' the soil, to dig compost or other fertiliser into the soil, to cull some weeds. In the short term, this can appear low impact. But in the long term, the damage is profound. It's a long, slow drip, a leaching of available nutrients, where the soil doesn't forget, but it can happen too slowly for us to notice in a hurry.

For a long time, the plough was seen as noble, and necessary. Thanks to technology, we have ploughed much more than ever over the last century. The advent of the internal combustion engine, cheap fuel, cheaper fertiliser, and all kinds of gadgets for the backs of tractors have led to massively increased use of the plough over way more of the Earth's surface. In any one year, half of the world's arable land, the bit that grows crops, is left bare – usually from the plough.[7]

Continuous ploughing, especially interspersed with sowing a single species of crop, is the best way to bugger up soil. Soil simply doesn't have the physical or biological ability to cope. If you're losing soil 100 times faster than it takes nature to make it[8] – losing carbon, losing structure, losing soil life – then it doesn't take a genius to see that we're in for some trouble when it comes to feeding the world.

The good news is that lots of farmers are changing.

Ploughing rates are starting to drop in many places, as no-till agriculture proves its worth.

No-till agriculture, developed in a few cultures – but mechanised for big agriculture in Australia, and now used widely in the United States as well – is a method of inserting seeds straight into the ground in a hole, without the need for digging soil. To suppress weeds, the land is usually sprayed with a herbicide first, most often glyphosate.

No-till agriculture has been a great boon to soil, helping reduce erosion rates and improve carbon storage. Curiously, however, in the United States, where no-till has a strong take-up rate, at about 50 per cent of grain growers,[9] the majority of farmers will still put the plough over the paddocks every few years.

Unfortunately, no-till is still damaging, however. The International Panel on Climate Change estimates that soil is still lost up to 20 times faster than it is replaced in no-till agriculture.[10] That's admittedly about a fifth of the rate of loss with ploughing, but it's still a backwards game. The use of herbicides, the lack of animals and the planting of single species are all possibly to blame.

Along with no-till, artificial fertiliser use is becoming, in some areas, more strategic, less wasteful, and more targeted. That's a good thing for the farmer, and the farmed. A good thing for soil, and for the environment further afield.

In fact, some of the best modern ways to grow grains are in pasture, and to also use grazing animals. Colin Seis, an Australian sheep farmer, has led the charge. He's perfected a system of direct-drilling grain seeds into pasture. Through a complex series of cropping, resting and grazing, he's able to grow food and store carbon and build soil life. He does all this without resorting to glyphosate, while keeping living plants in the ground all year around.

On his New South Wales property 'Winona', Seis has seen soil nutrient availability boosted 172 per cent on average, and his carbon by 200 per cent.[11] He reckons a lot of farmers still work on what he calls the 'Moron Principle', where they keep putting more things on. More nitrogen, more herbicides, more phosphorus.

Through trial and error, Seis has led a global movement to rethink how we grow food. 'Pasture cropping', as it's called now, is practised by over 3000 landholders on 3 million acres around the world.[12] The aim is to cut out inputs, and concentrate on soil health to grow crops. The aim is living soil, healthy crops *and* a living wage.

Modern farming has other, more insidious impacts. So much of what we do in modern farming has sped up soil's oblivion. On balance, some practices are bad for soil. Ploughing is bad. Adding artificial nitrogen is bad. Ditto for adding phosphorus. But so, too, is the rampant use of herbicides and insecticides.

There are now crops that don't die when sprayed with herbicide, because they've been genetically modified to resist them. It seems like a brilliant idea, because farmers can spray the crop, and kill only the weeds (though herbicide-resistant super weeds are a problem).

Since the advent of corn, soy and cotton that are resistant to glyphosate (Roundup), the use of herbicides has reached strato-spheric quantities – with around 8.6 billion kilograms (19 billion pounds) of the herbicide sprayed around the world.[13] The use of glyphosate in the United States – where grain growing is subsidised, and is some of the most intensive on the planet – has risen 300-fold since 1974, and 15-fold since 1996.[14] A 2016 paper in *Environmental Sciences Europe* estimates that between one-quarter and one-third of all cropland, globally, is sprayed with glyphosate every year.[15]

The problem with glyphosate, of course, is that it kills not only the 'weeds' we see, but the subterranean life we don't see. Some is passed directly down into the soil as root exudates, immediately having an impact on the rhizospheres. When every plant in a paddock dies, you also lose the entire first trophic ('feeding') level – the initial sugars that feed the whole system – because there's no photosynthesis. And in the absence of fresh sugars from plants, much of the microbial life dies. Fungal populations that are reliant on their association with roots are particularly badly hit. Glypshosate actively compromises fungal colonisation of soil, and blocks the synthesis of essential amino acids by soil microbes. It lowers plant immunity. It's also been suggested that glyphosate can lead to more plant disease, because its use 'reduces the overall growth and vigor of the plants, modifying soil microflora that affects the availability of nutrients required for disease resistance'.[16]

In other words, kill the plant, kill the soil – or at least mortally wound it.

Glyphosate use on genetically engineered 'Roundup Ready' crops (like corn – for ethanol perhaps?) means there are still living roots intact in the soil, which is a good thing. But the plants also exude some glyphosate into the soil, which isn't a good thing. It affects the growth of fungi, and alters the soil's biome.

So no-till agriculture, where the land isn't ploughed, but seeds are direct-drilled into the now-dead, glyphosate-treated fields, doesn't have clean hands. Yes, it reduces carbon emissions and erosion compared to ploughing, but by killing life on top and within the soil, you've diminished the underground community, released soil life carbon and soil organic carbon through the action of the weedkiller, and cut off the deposition of glomalin.

All of this has consequences on soil. In other words, no-till agriculture that uses herbicides (and not all does, such as Colin Seis' method) means we're still going backwards – just not quite as fast.

It's not just broad-spectrum herbicides that deplete soil, so do other tools in the modern farmer's armoury. Fungicides, which are widely used to discourage mildews and moulds on crops, can depress all those other wonderful fungi that are feeding plants and gluing soil together. In fact, fungicides, herbicides and soil fumigants usually sound a death knell for soil algae, too.

What about insecticides? Surely, they don't kill soil?

Well, remember all those little lives we met? The worms and dung beetles? The springtails and arthropods? Insecticides are virtually always broad-acting, and indiscriminate, meaning they kill everything, or at least a lot of different things. You might want to get rid of aphids, but you may also kill harmless native wasps – perhaps the only ones that can fertilise a native tree. You may spray an insecticide, say one called Steward, for moths, and take out the local bee population at the same time.

And speaking of bees, around the globe, bee populations are crashing, in part because of pesticides. It's not just crops themselves that are affected by bee deaths, or the farmland. Every time a parkland or native forest can't fertilise itself, soil loses.

Neonicotinoids, the most commonly used class of insecticides, are thought to have sub-lethal effects on bees, but are increasingly being implicated in hive disorders and collapse.[17] They can also have an unintended effect on crop yields, by taking out predators that would otherwise keep pests at bay.

Below-ground creatures are also affected. Worming agents, used on livestock, not only kill the intestinal parasites that they're intended for, but also all the other things that call poo home for part of their lives. Forget 7 million worms per hectare – throw a bit of worming agent on your cattle and you could end up with next to none.

When I think of soil, I think of it as a living, breathing, complex organism. And like my own body, I think of all the things that it's designed to do. And just like my body, everything I do has consequences. Cut my skin, I bleed, and am more open to infection. Cut it again and again and I scar. Sever the source of food to one part, the blood supply, and that part dies, like frost bite. Feed me badly and my immune system is weakened. When I take a therapeutic drug, I'm essentially aiming to alter my body's metabolism or activity in some way. It's interfering with normal body processes. That could be worth doing, of course, to help me fight illness or disease, or to suppress pain. But interfere is the point. There may be side effects, and the way to judge a drug's use is whether the side effects are worth it.

The exact same thing is true with insecticides and herbicides. Their whole aim is to interfere with normal biological functions. They are disrupters by nature, and *of* nature. In the same way that I only take antibiotics when I need help beyond my body's own defences to fight infection, soil, and the life it supports, can only be negatively impacted by pesticides and herbicides unless it's a time of crisis, rather than a normal season. The whole point of them is to kill life (the clue is in the last bit of the name, –*cide*, meaning something that kills) – and if you use pesticides and herbicides, you have to know that there will be a reaction. To think otherwise is to misunderstand nature.

Growing food is about harnessing the goodness of our land. Much of what we've done, historically, is about killing, not letting things thrive. It seemed right at the time. But we do need to be honest with ourselves about what we've done right, what we've done badly, and what we can do better.

Every time we grow food, something else wants to outcompete it or eat it. That's the whole reality of life. We can interrupt biological functions to get the food we want, by ploughing, adding artificial fertiliser, using pesticides. But all of these have a cost. Those interruptions affect the air and the waterways, and while the broader environment also suffers, most of the cost is actually borne by soil. Insect communities can rebuild, if left alone enough. Plants can recolonise areas. But soil damage can be long term, because it's structural, biological *and* chemical. It's a three-dimensional ecosystem.

Research has shown that we're losing the battle. Despite our armoury of chemicals, and the multinational companies involved in the agricultural industry, 95 per cent of herbicides and over 98 per cent of insecticides are not even reaching the targeted pest.[18]

Use a mercury-based fungicide and the soil accumulates the heavy metal, which can then be passed on in food. Just because a chemical doesn't wash into the waterways, doesn't mean it hasn't been absorbed into soil. Fungicides with a half-life of a few days have been found in soil after 100 days. Glyphosate half-life in a field is estimated to be 47 days, but has been found tightly bound to soil a whole year later.[19] A full 45 per cent of European agricultural land tested contained glyphosate, or its main metabolite, according to research that came out in 2018.[20] This stuff is everywhere, all the time.

You can't see all the damage done to our soils by the plough, and by pesticide and fertiliser use – and historically, the damage has been hard to measure. But now we know that damage is caused at the microbial level.

The thing is, soil doesn't forget. What we do to soil, we do to ourselves.

SAVE OUR SOIL

The best way for growers to protect soil is to avoid bare earth – whether it's in your home garden, or on the world's biggest farm.

Don't dig if you can help it. Ideally, keep continuous, living plant cover if possible, mulch any bare earth, and encourage diversity. The more diverse array of plants you can grow, the more diverse your soil microbes become, and the more resilient your land is.

CHAPTER 13

Burying Charcoal and Building Soil

Wim Sombroek was only 10 years old when the full impacts of the Second World War hit his homeland's food supply. In 1944, The Netherlands was ravaged by the *hongerwinter*, the hungry winter – a famine brought about by years of war. Wim watched as his father worked miracles, feeding his family from a small plot of land, fertilised over generations, including with ash and half-burned coals from the fire.

It was watching the black, so-called *plaggen* soil of his family plot bring forth food that gave Sombroek an affinity for black earth. A typical soil scientist (his profile picture is usually of him with a neat moustache and parted hair, in wire-rimmed glasses, looking a bit like Ned Flanders from *The Simpsons*), he was Secretary-General of the International Society of Soil Science from 1978 to 1990. But it was growing up with the human-induced fertility of soil that made Sombroek very aware of its potential to change lives. It is Wim Sombroek who we have to thank for a whole new way of looking at dirt.

In Brazil in the late 1950s and early 1960s, Sombroek came across dark, impossibly fertile soils in the Amazon Basin. Feted and praised by locals, these rich soils made a wondrous growing medium. Sombroek studied them for his PhD thesis, and in 1966 published a book, *Amazon Soils*,[1] initiating scientific exploration into these unnaturally dark soils, which became known as *terra preta*.

It turns out *terra preta* – literally, 'black soil' in Portuguese – is anthropogenic in origin, the result of human intervention over many centuries.

Most of the Amazon isn't actually as fertile as many people think. In fact, overall, it's a pretty poor place to grow crops. Thanks to nutrient-leaching from high rainfall, and the fast rate of organic carbon decomposition due to the warm temperatures, Amazonian soils are generally not very suitable for agriculture.

However, certain parts of the Amazon Basin have far richer soil than the relatively impoverished soils nearby, a difference that can't be explained by geology. These patches – many about 20 hectares in size, but some up to 350 hectares[2] – have been made astoundingly fertile by the deliberate actions of people.

What's more, these *terra preta* patches have been fertile for a thousand years, some up to 7000 years.[3]

Far from humans always mining and eroding soil, this shows we can also build it.

Millennia ago, some Amazonians worked out that putting things in soil makes it more fertile. We don't know how they knew. We don't even know how they did it, because these farms were abandoned after the original inhabitants died, probably of smallpox, after Europeans arrived in the 1500s and 1600s. But we can guess. One of the things you always find in *terra preta* is charcoal. Yes, half-burnt wood. *Terra preta*, sometimes called Amazonian Dark Earth, has 70 times more charcoal in it than comparable soils nearby.[4]

The charcoal in *terra preta* is often called biochar. Essentially it's wood burned in a low-oxygen environment until it is porous, mostly carbon, and stable. The Amazonians buried the charcoal – probably with animal and possibly human waste, and potentially food scraps.

Terra preta soil is, quite often, 13–14 per cent organic matter.[5] That's five to ten times as high as most agricultural soil in Australia.[6]

So, how much more fertile is it? *Terra preta* contains up to eight times more carbon and nitrogen than adjacent soils, and has 1000 times more available phosphorus.[7] On *terra preta*, you can grow twice as much food in the same space, with the same labour, than you can on nearby soil that isn't black.[8]

From many accounts, you can grow a crop for 40 years on *terra preta*, without adding anything, without losing fertility.[9]

Growing crops in the nearby soil using slash-and-burn agriculture, where you cut the trees down, burn them and turn it into cropland, can be done, in theory, sustainably. But – and this is the *big but* – you can only plant crops on such soil for between one and three years, and then it needs to be left fallow, to rest, for between five and 25 years to replenish.[10] This isn't a good model for feeding a lot of people for a long time.

Terra preta, however, is. These fertile patches fed relatively large populations over a very long time. There were up to 100,000 people in some pre-Columbian settlements.[11]

So how does *terra preta* work? I could go into cation-exchange theories, and slow oxidation leading to carboxylic groups. But really, it seems that carbon, in the form of charcoal, can store minerals in a manner that plants can take up, but water can't flush out. It's porous, so it holds air, too. It's a safe home for microbes, and the hyphae from fungi can wheedle their way through it.

Its porous nature also means that it has an incredible surface area. A single gram of properly made charcoal can have a surface area of up to 2000 square metres – the equivalent of seven and a half tennis courts.[12] Put another way, 10 grams would have enough surface area to cover the oval at the Melbourne Cricket Ground.

Terra preta is only one example of the ways humans have made soil. We've been doing it for thousands of years, on most continents, it seems.

Those *plaggen* soils that Wim Sombroek's father worked in Europe were developed at least as long ago as the Middle Ages, perhaps even as far back as the Bronze Age.

Plaggen – meaning sods, or divots, or turf – takes its name from the way it is made. To care for livestock in the cold European winters, people made bedding for the animals from clods of grass (often including roots), cut heath, or even the top layers of fallen leaves from the ground, along with some humus – soil organic matter. The animals lay at rest, doing what animals do, like going to the toilet and shedding hair.

At the end of the winter, the animal bedding was spread over paddocks. Over the years, this rich addition built up soil to a depth of over a metre and a half, where it would've been just 30 centimetres deep without *plaggen*.[13]

The fertility of much of Europe isn't natural. It was made by people.

A form of this *plaggen* was also carried out on remote islands, such as the Shetlands – even until the 1960s. Sadly, most went the way of the dodo (and the worm, and much of our topsoil, actually), when artificial fertilisers were introduced, early to mid last century.

Humans have been making soil, on and off, for at least 7000 years. Yes, we've made mistakes, but different cultures in South America, New Zealand, Europe and Asia have worked out ways to either conserve soil, or make more of it.

In Asia, the terraced field is brilliant for storing carbon. In China, it's been shown that terracing can increase carbon storage by 34 per cent.[14]

Africa, too, was doing quite well until Europeans arrived.

In western Africa, terracing, crop rotation, green manuring and mixed farming were common prior to European colonisation;[15] paying attention to soil allowed empires to be built and sustained.

It's now starting to be recognised that African nations such as Liberia, Ghana, Benin and Guinea also have very dark soils, not dissimilar to *terra preta*. Once thought to be natural, they, too, are actually the result of human efforts. Some have been dated to be 700 years old.[16] Independently of the Amazonians, agricultural soils were being enriched with charcoal – increasing soil organic carbon by up to 300 per cent, and at the same time creating the conditions for twice as much plant-available nitrogen, and an astonishing 270 times more plant-available phosphorus.[17]

African Dark Earths, as they are known, might not achieve quite the same scale – in terms of amounts of land, or of fertility – as Amazonian Dark Earths, but by modern farming standards they really do look like a gold standard.

The whole world could learn a lot about soil health from Liberian villagers from 500 years ago.

Another European mechanism for making soil – one that seems to only be practised now by the hippies and permaculture junkies – is *hugelkultur*, a German word meaning 'hill culture'. Anecdotally, this soil-building technique was born of the forestry industry in centuries past. Waste wood from logging is laid around hills, perhaps mimicking terraces, or put into shallow trenches and piled up, then covered with earth. These mounds are about 30 per cent wood and 70 per cent soil, with no wood showing, only the earth visible on top. Hence the reference to 'hill'.

Hugelkultur is simply composting at work. The timber provides a cracking environment for white-rot fungi to work in. The decomposition warms the earth a little as the wood rots, allowing for a slightly longer growing season in temperate climes. As the wood is slowly consumed, it is turned into soil through the action of microbes. The structure allows for air pockets to form, which is better for air and water infiltration, and worm and arthropod movement, along with root penetration, too.

Hugelkultur is one of the few efficient ways to turn limb wood into soil, without using fossil fuels to chip them or burn them first. In the

United States, it's estimated that 13.3 million tons (12 million tonnes) of backyard trimmings – leaves, grass clippings, tree branches and the like – are put into landfill every year.[18] This, in theory at least, could be turned into topsoil.

In Australia, we don't practise *hugelkultur* in forestry. We simply burn the wood, releasing all its carbon instantly, killing microbial populations on and in the ground nearby, and soiling the sky. Yes, our wood is harder. And yes, we do have a drier climate, which means the breakdown of the timber might take longer. But imagine if we reconsidered that timber, and thought of branches and crooked logs as a resource?

Imagine if, instead of clear-felling a logging coupe, scorching the earth with fire, then leaving it bare to erode, we dug shallow trenches and buried the 'waste', covered it with soil, and left it to create fertile patches. Imagine if we did this on the contours, creating ten-metre (thirty-foot) long plateaus to slow down water – mounds that would produce thick, rich soil for decades to come.

Imagine if we set fire to the mounds, a low-temperature fire, and *then* covered them with soil so it created charcoal in parts, that was also buried.

Would it work on an industrial scale? Well, nobody seems to be trying to find out.

There's precious little research into *hugelkultur*, and virtually none that is peer reviewed looking at nutrient retention and soil-building capacity. That doesn't mean it won't work. It just means we haven't had the imagination to test it out yet. It's solely the realm of the 'fringe' elements – which, it must be said, have got a lot of other things right about soil that science now backs up.

Biochar is gaining some traction, thanks to the role of *terra preta* in proving soil fertility and carbon sequestration. Biochar is the kind of charcoal that benefits soil. It's a bit more scientific than just setting fire to a bunch of logs, then burying them while they're still glowing. But not impossibly so. The thing is, it's full of promise.

It's been shown that you can burn fossil fuel coal to produce biochar, and still be carbon negative.[19] Carbon that can sit in the ground, aiding fertility, for a thousand years or so. There are plenty of ways to make it at home, if you search the internet.

As we've seen, the Green Revolution has depleted soil. It has, inadvertently, sucked carbon from the ground and eroded agricultural land. It has aided and abetted the industrialisation of the animal industry, and is now responsible for producing bio-fuels of dubious benefit to soil and the climate, in the long term.

Imagine if we'd had a Black Revolution, where Dark Earths were celebrated, biochar lauded, and long-term fertility the aim.

I think Norman Borlaug's motives were noble. Unlike Fritz Haber, who knew exactly what he was doing when he invented poison gases and debased the Nobel Prize, Borlaug thought only of how to feed the world, the best way he knew how. But he did spend a long time ignoring some science that didn't fit his worldview – that ancient, fertile soils, made by humans, were resilient, productive and long lasting. It wasn't just Borlaug, of course, it was lots of people in power, in universities, in research laboratories, who focused on yield, while ignoring just what it *is* that does all the growing, and how we need to nurture the medium that gives a plant its best life.

Growing food is more complex than simple genetics, and N, P and K. It's more complex than the Green Revolution would have you believe. The answer was under our feet (well, under the feet of some Amazonians and Africans) the whole time. We just didn't look there.

Science takes a while to find the answers. But backyarders, permaculturalists, market gardeners and ancient tribes often have wisdom that science takes a while to understand, or appreciate. We have ignored Dark Earths, and the potential for a Black Revolution for too long. We have focused on the highest-yielding plants, and ignored the best soil.

The best soil, these days, is probably not in the Amazon. It has probably not been made by nature. It's certainly not on an industrial farm, or even on a commercial farm, like ours in southern Tasmania.

The best soil, growing the best food, is the soil that has been given the utmost attention – most likely by the person who is going to eat that food. It's in the home or urban garden, where the effort you put into your soil can not only be felt in your hands and back, but also be tasted on the plate.

The best soil is waiting for you, or waiting for you to make it, outside your back door.

CHAPTER 14

Weeds: What We Can See Tells Us About What We Can't

Peter De Vries is showing me the market garden he farms with his wife Prue, tucked under the buttress of an old eucalypt forest. Living where I do now, in the southernmost shire of Tasmania, an island off the southern coast of Australia, can feel like I'm a long way from my urban roots. But I'm amazed I can still be heading west from our farm and not sink into a peat bog, meet a drop bear, or fall off the edge of the world. Peter and Prue live a long way from anything resembling suburbia, or a town. As I drive, I see rainclouds looming ominously over the remnants of Gondwanaland, the World Heritage-listed South West Wilderness.

In this corner of Tasmania, nestled hard up against ancient rainforests, there are trees still growing that were around when Jesus was a lad.[1] The weather belts in like a Mack truck, and can take all with it. A hill, a forest, both to the west, protect Peter's greenhouses from destructive winds – gales called the Roaring Forties, after their latitude and attitude.

I'm here to talk greenhouses, but Peter De Vries wants to show me weeds.

'You can tell what's going on with your soil by the types of weeds you grow,' Peter exclaims.

'Really?' I say, unconvinced. 'Do you mind showing me how you built your greenhouse…?'

Fast-forward a decade and I'm pretty disappointed that I didn't pay Peter more heed. I put his talk about weeds down to woo-woo, the mystical part that some growers seem to rate. That talk in gardening circles around planting by Moon cycles, filling cow horns with dung and never sowing seeds with your right hand (okay, so I made this last one up). To someone new to growing, the ability of weeds to signify soil deficiencies seemed just a bit too easy and mystical to be true.

To my shame, I now know much better.

Of course a weed tells you a bunch of things about the soil underneath. A weed is just an opportunistic plant, after all. Everything is connected, and the visible tells us about the invisible.

Pioneer species – plants that arrive first in disturbed soil, like thistles – help signify that the soil has been laid bare or dug over.

Dock, a broad-leafed, tannic-flavoured, deep-rooted plant, thrives where the soil is dense, where it's a bit airless and compacted. It might grow in the laneways of a former dairy farm, or the rows between trees in an orchard where the tractor has been.

Sheep sorrel, an edible, lemony flavoured weed, is usually indicative of soil that is a little acidic.

Bracken likes well-drained soil, and is partial to particularly low pH (high acidity). Clover can indicate low nitrogen. Queen Anne's lace is a sign of low fertility, too. Capeweed tends to show up in more sandy, well-drained soils – perhaps those not lacking in phosphorus.

A weed is simply a plant in the wrong place. Weeds are not bad in or of themselves, but rather can be a useful tell-tale symptom of soil. Despite being treated as inherently wicked, lots of them are doing great things. In compacted soil, 'weeds' like dock or dandelion are more likely to take hold. Both are blessed with long, fat, fibrous roots that spear down vertically and can access water more readily in summer, the plant often able to remain green when all the grasses either side are browning off. They break up the soil as they descend, and can store organic matter lower in soil. They are soil conditioners, really, helping to break up the earth and allow in more air, while drawing up minerals held deeper down. They may linger for a year or three, but when the soil changes, so do the species that inhabit it.

Plants colonise an area, in a process called succession. If the ground is bare, the first plants that grow can cope with that. They like those conditions, and they help heal the earth's open wound. But they also alter the ecosystem to favour other plants.

In our part of the world, the logical successions would lead to a forest, because that's what grew here. So thistles or dock might be followed by bracken, which could well be followed by grasses, then the introduced blackberry, then wattle, which shade out the black-berries, then gum trees. Each plant arrives because the conditions are ripe for it to thrive.

Plantain, for instance, also has a decent tap root. This 'weed' estab-lishes itself quickly in gravel or between pavers, as well as on parts of our paddocks after the pigs have dug them over. Most of you might have seen varieties of plantain, the narrow-leaf species, near a driveway or footpath. It has long, pointed leaves, wider than grass, with marked ribs that run along to the tip. It puts out brown seed heads that look like witch's hats fringed with a daisy chain of tiny flowers. They colonise an area, putting down roots to help break up the rock, holding carbon and drawing it down, becoming high in calcium and iron themselves in the process. They might be a bugger in your pavers and driveway, but they are a boon to help heal land for growing food, if you have time to let them do their thing.

Another example of a 'weed' that changes its environment is clover, which tends to colonise areas low in nitrogen. The clover – through its microbes, of course – is a nitrogen fixer, which it traps from the air. A lawn rich with nitrogen is less likely to support much clover compared to one that isn't, and so clover can eventually do itself out of a job.

Weeds exist because they're healing soil. They're colonising your garden, cropland or paddock because there's a vacancy in the eco-system for them. They're there, persistent and resilient, because the conditions in the soil – both its life and its structure – are ideal for the weeds, and not necessarily ideal for what we want to grow.

Importantly, in the process, the weeds are feeding subterranean communities, and often fixing the very problems in soil that have allowed them to thrive in the first place.

The reason weeds are there, competing with our crops for water, nutrients and space, is because the soil is more amenable to them than our domesticated plantings. Often, on our arable land, we try to grow things that want to die, and try to kill things that want to live.

If weeds are just an indicator of what is going on in and under soil, so too are other things we can see. In fact, for the home gardener, or even the smallholding farmer, visual clues are vital. It can be as obvious as a monoculture. Nature doesn't like single species. Plants like to live in a diverse colony. Name the old forest or natural wetland you've seen that allowed only one species of plant to live. A variety of species, and of ages, is fairly normal in a thriving ecosystem, and what happens above ground is a really good way to see what is happening below. You don't need an electron microscope and DNA analyses for this; you just need to pay attention to what grows where, and when.

Work called the Jena Experiment in Germany backs this up. Scientists planted crops in single species, and multiple species, with up to 16 species in a plot.[2] The more species, the healthier the crop. The more species above ground, the more species of microbes, fungi, arthropods and mites there are below ground. Instead of outcompeting each other, the plants actually work in mutually beneficial ways. They are taller, more pest and disease resistant, more drought tolerant, and have greater leaf area. And of course, they don't do this on their own. They need healthy soil.

In healthy soil, abundant species of microbes do a thing called quorum sensing. Now, this gets a bit complicated, but bear with me for a moment.

Bacteria can count. Yes, I know they're single-cell organisms, without a brain or central nervous system. But they can sense when there are other bacteria like them nearby – and also bacteria that are not like them, and how many there are. They also communicate with each other. A good example of this in action is one of the most astonishing, because it's visually stunning.

Some creatures, such as certain squid, are bioluminescent. They glow. Actually, *they* don't glow, their microbes do. When there are enough of these microbes, and conditions are right, all the microbes put out a chemical that allows the squid to shine in the dark. It's like a chemical message, or a chemical vote, if you like. When enough microbes 'vote' in a certain way, they all start work.

Hawaiian bobtail squid are a prime example. They live in shallow waters, only a couple of feet deep, and nestle in the sand during the day. At night, they venture out and have this incredible ability to sense the amount of light from the night sky – and then glow in the same way, which stops them casting a shadow. But it's not the squid that's glowing, it's a bacterium called *Vibrio fischeri* that the squid has taken into its body, and which it has a symbiotic relationship with. It uses the bacteria to make it next to invisible, to protect it from predators, and to make it easier to find prey. The thing is, *V. fischeri* has to be in high enough numbers to glow. Too few of them, and they remain dark. But herein lies a problem. The squid needs to glow in just the right way; it doesn't want to come out the next night shining like a light bulb. So every morning before bed, it ejects most of its bioluminescent bacteria and goes dark. The bacteria then breed up again during the day, and when there are enough of them, they light up again just before the squid heads back out to hunt.

This amazing ability of microbes to know how many other bacteria are present has been termed quorum sensing. The term quorum most often applies to a meeting, or vote; you need a minimum number of people in the meeting to make decisions, and this number is a quorum.

Same with microbes. If there's not enough of them, they lie idle – but, get enough of them together in numbers (and perhaps in species, depending on the time and place), and they will fire into life. To do this, to 'count', they put out chemical signals, which tell all the other bacteria who is there. A bacterium is constantly getting information on precisely what other bacteria of the same species are nearby, and who else could be around.

It may be easier to imagine it like this.

When a species of bacterium, a disease-causing one, enters our body, it probably won't make us sick, because we have very good defences against bacteria of all kinds. So it enters our body and goes very quiet. It tries to hide its presence while it increases in number, until there are enough bacteria to make an attack. Kind of like a Trojan horse full of breeders. As they hit quorum, the bacteria all act in unison to perform a pre-determined action. In a health sense, that could mean a pathogenic bacteria making us sick.

When quorum is reached in soil microbes, it's exactly the same. Some of the microbes in the soil – a lot it seems – are only triggered by high numbers. Too few and they don't start work. But get them in high enough populations and our subterranean communities leap into full action, cycling nutrients up to plants, and between plants. It's not as striking to the eye as luminescence, sadly for soil, but it's the same principle. Quorum sensing is just as much a part of soil health as lighting up is for the bobtail squid.

With plants, quorum sensing means that instead of competing for nutrients, they can help provide them for one another for mutual benefit. But only in healthy soil. You need a complex array of living plants, all feeding a complex array of living microbes, to start the process.

The reason what we see on top of the soil is so helpful is because what happens underground is so hard to imagine, and really hard to measure. Despite advances in science, what's happening in each patch of dirt is hard to gauge without a lot of time and money going into it. Of course it's possible to do accurate soil analyses, if you have a lab full of staff, and can conduct endless tests and DNA scrutiny of the soil microbes. But most of us, from professional growers to those who just want some parsley on their pasta, have to rely on tools more readily to hand.

Most important is sight.

As a farmer, I look for lots of things. Soil cover, the nature of the ground underfoot, the species in the paddocks, the variety and succession of plants.

As we've seen, bare soil is damaged land. Every time you see bare earth, that is soil dying, whether that's in the pig paddocks or the market garden. It's amazing the damage we do to grow vegetables.

To see how the farm is doing as a whole, we check the colour of the creek. A clear-flowing creek is a good sign. A brown creek means we're losing topsoil. It's a simple, but effective, way to judge erosion.

Anywhere that we see just one species, we know soil is being starved. Nature doesn't do it; plants don't like it. A monoculture crop is the worst. The endless fields and fields of canola or corn or wheat that so many countries now boast are a sign of soil gasping. It's even worse if these crops aren't rotated, with different crops grown in between.

But it's not just monoculture crops. It's important to look for diversity in the paddocks. Diversity in the grasses and broad-leafed herbs and trees.

And it's not just diversity in plants. On our farm we want more insect varieties, and lots of insects in total, even as we try to manage the pest species in the garden. Ecological harmony above ground equates to ecological complexity and harmony underneath.

To judge soil health and life, we also look for changes over time. Areas of the farm or garden that brown off quicker in summer, or slow down their growth more quickly in winter, are indicative of problems in soil. Healthy soil holds water better, so it's more resilient in the hotter, drier months. Because soil life (and to a different extent, soil carbon) actually has its own ability to change the temperature of soil, if one area slows down its growth quicker in the cold months, then we can assume its microbial life is more subdued.

These days, I look at soil differently, wherever I am. When I pass a farm that has sprayed glyphosate or other herbicides over broad acres, I know the soil life is arrested. If I can't find a layer of half-composted leaf litter underneath the crowns of grass in my paddocks, I know my soil is going hungrier than it should. If I find thistles thriving in an area that hasn't been bare in the last couple of years, I know my cattle (or the wallabies) must've overgrazed it. Deep-rooted, more broad-leafed grasses outcompete thistles. But grass that is grazed too often reverts to finer-leaved species with shorter roots. Roots that can't access water as late in summer as the ones they replaced. Grasses that can't outcompete the deep-rooted thistle, that die off in the first big dry.

We look for insect and bird life. We listen for frogs. We are conscious of pugging, where an animal has left deep footprints in the earth (or a naughty visitor has done the same on a vegetable bed in the garden). We note the shade of green in the grass in various places.

And of course, Peter, I do notice the 'weeds' – though I now think of them more as just another form of life to manage.

What happens in your vegetable patch might seem disconnected from the gardens around you. How you treat your soil might seem arbitrary when seen in isolation. But what you do, how you manage your patch, how the food you eat is grown, has an impact way beyond your fence line.

Remember how the fungal hyphae growing around the roots of your plants can travel 2000 times further than the roots themselves? Your neighbour's garden affects yours, as yours affects theirs.

What we can see above ground tells us a lot about what we can't see. It doesn't tell us everything, but it does give a window into the magical life underneath our feet.

And what's under our feet can affect our mood, the rainfall, our own microbiome, the global climate and our futures – both on a personal scale, and a generational one.

What can you do to change the trajectory? Choose food from better farmers, for a start. Or grow your own.

What's the best thing you can do with the little patch of the world that you *can* control, because you work it with your own hands? What can you do in your garden?

Make compost. And perhaps plant charcoal. But most importantly, think of yourself as a grower of soil.

CHAPTER 15

Home Gardeners Rock

'I'm growing carrots,' the young woman beams.

'Really?' I say, trying to sound enthusiastic. I'm on a road trip, doing lots of public appearances, and the sheer joy I find in so many wanna-be farmers is captivating, but exhausting.

'Where's your farm?' I ask, hopefully.

'I don't have a farm yet, but one day …' she grins.

'How big is your garden?' I ask.

'Oh, I don't have a garden,' she says. 'I live in a flat.'

'Really?' I say, even more interested now. 'Do you garden on your balcony, then?'

'No,' she says. 'I don't have a balcony. I've got a polystyrene box, though.'

'How does that work?' I ask, genuinely puzzled.

'Well, I get a bit of sun coming through the kitchen window,' she explains. 'And I sowed carrots in some good soil in the box, and it sits on my kitchen table. I move it around to catch the sun during the day.'

'And you can grow carrots that way?' I ask, stunned.

'Yes!' she grins. 'And they're amazing.'

This woman, whose name I never knew, inspires me. That she can take proper soil, a skerrick of light, an otherwise environmental catastrophe like polystyrene, and turn it into food – in an apartment no less – is astonishing.

We can grow food on our small farm, enough to make about 5000 feast meals annually, and others do amazing things in their own spaces.

But I never thought you could grow carrots on a kitchen table. Inside.

The thing is, it's inspiring because looking after soil is something we can all do, if we rethink how we grow and eat.

For my entire adult life, I've been around food in more than the usual, fuel-yourself kind of way. I worked in a bakery and delivered milk before I left school. I went straight from year 12 to a chef's apprenticeship. Even when I was at university, I worked a few shifts in restaurants, and cooked a helluva lot at home. And when I discovered restaurants as a diner, not as their cook, I thought I'd found nirvana. I went from eating quantity, to quality.

As a restaurant critic, it wasn't Sydney's three-hatted Marque, or Britain's Gordon Ramsay, or even Paris' Pierre Gagnaire that made me really take note. It was a lettuce, eaten in a home gardener's domestic kitchen in the suburbs of Canberra, the town I grew up in.

It was there, eating lunch, watching the gardener patiently scoop tomato seeds from his sandwich and lay them in neat rows on used envelopes, drying them ready to sow for next year's crop, that I suddenly realised what I'd been missing: home-grown food. Something grown out the back door. The kind of food we humans have, for most of the last 12,000 years, grown as our staple.

The fresh, nutrient-dense, flavour-packed food that we've always had as humans, until the last 100 years or so.

What this lettuce made me realise was that it's not only the leaf that matters, but how that leaf was grown. Sweet, complex, immaculately fresh, and yet not overly large, it was a thing of beauty, gastronomically. It was a prime example of what people are snipping from their garden beds or pots and putting on lunch all over the country – but only if they grow it themselves.

It was the ultimate expression of how food can, and should, taste, when grown with care. The care that I could sense, I now realise, was in the gardener's treatment of his soil.

The great news is that that soil, lovingly cared for soil, with all its attendant microbial life that we met earlier in the book – that can be yours, too. And while it is full of complexity, it doesn't have to be complicated.

Growing food, it seems, has become overly complicated. Commercially, it is not just the result of a thousand industrially made chemicals, but also vast distances. Estimates put it that each calorie of food we eat in places like Australia is the result of 10 calories of burnt fossil fuels used to grow the food[1] – releasing carbon not only from the fertiliser and the tractor, but also from the soil. That's even worse if the food comes from a long way away. In the United States, the average foodstuff on a supermarket shelf has travelled an average of 2400 kilometres (1500 miles). In a fossil-fuelled truck.[2] The average basket of food in Australia has travelled about 21,000 kilometres[3] (13,000 miles) – a distance that would almost encircle the entire continent.

For many of us, however, this doesn't even enter our psyche.

Australia loves to think of itself as the breadbasket of Asia. As a country that exports vast amounts of food to those in need. Rough estimates suggest we grow enough food here to feed about 60 million people[4] – more than twice our population. With only about 1 per cent of us doing the growing, we are held up as an example of a nation that punches well above its weight when it comes to putting meals on the table.

In 2012, I went to a discussion of the future of food in Australia. Rarely was it ever thought that we should talk about food sovereignty; the ability of Australia to grow enough of its own food for maximum health, even in times of trouble (a scenario the COVID-19 pandemic of 2020, it must be said, brought into stark relief for many us).

Even less important has been how nutrient dense that food is. Just grow lots. Export lots. Bigger is better. More means productivity and profitability.

We have no overarching plan for how to grow all the necessary food for ourselves because we're a net exporter. Seen as a whole, this nation *is* growing plenty for us, and then some. Problem is, questions need to be asked. Where is this food grown, how is it grown? Is it the best use of land, and is it the best food we can access? Have we mined soil, or improved it? Have we grown nutrient-poor food in poor soil, using artificial fertilisers to pump it up in volume, but decreased the value of our land – our national asset – while in the process growing gastronomically hollow foods? It's a question all nations need to ask.

We know that some of the worst food availability in Australia is in the regions, the supposed 'food basket' postcodes. Farming regions have worse diets, and worse health outcomes, than most urban areas. Remote communities, those really removed from the cities, struggle to even buy in something that has spent less than a fortnight out of the ground. Sure, we're a big country, about as big as continental United States (the bit without Alaska), and way bigger than Britain and Europe. It's about 4000 kilometres (2500 miles) from east to west, and 3200 kilometres (2000 miles) from north to south. Which means a lot of people live a long way from the source. And many remote communities are in hot, arid lands.

But growing food locally, it would seem, isn't even on our government's radar.

We all know fresh food tastes better. It's cheaper, more nutritious, and better for you if eaten in season.

And the best soil – from farms that have integrated all kinds of life into their ecosystem – are much easier to access if they're up the road, not on the opposite corner of the nation.

But you don't need to rely on farms for much of your diet, if you can look after the soil you've got.

On the other side of the globe, in a city built to great heights last century on the automobile industry, the gardens tell a different story.

That city is Detroit, the resilient industrial capital of Michigan where, for certain periods of history, urban living hasn't meant fresh food must come from a long way away.

In fact, over 23,000 Detroit residents are currently involved in urban gardening, and the city boasts more than 1500 individual urban gardens.[5] People might live in flats, and may not be growing carrots on their kitchen tables, but they still garden.

This is no accident. Born of the 1890s depression, Detroit's city gardens provide a lesson many other cities are taking in. The town is no newcomer to urban farming, however. Long before the car industry was even a thing, during the 'Silver Panic' of 1893 (the stock-market was new, and panics were common), Detroit's mayor, Hazen Pingree, watched as nearly 40 per cent of his population lost their jobs and came close to starving. He was adamant that residents would be better placed to cope if they planted their own food gardens. His efforts were initially mocked as 'Pingree Potato Patches' – a slight probably not unfamiliar to urban garden exponents today – but these urban gardens were a cracking success, and gave families more benefit than any welfare schemes of the time.

According to Kelly Vaseau-Sleiman, writing on the history of Detroit's urban gardens for Henry Ford College:

> Hazen Pingree knew that the city of Detroit had ... thousands of acres of land available in and around the city. Also, many (people) were desperate and willing to work. Pingree knew that many of the immigrants had farming experience. So, with the land and the willingness of the people, all he needed was the funds to start his program. Pingree took his vision to the wealthy, who chose not to help with the cause, calling the struggling working class 'lazy.' He then went to local churches, who scoffed at his plan, donating just under $14.00. Pingree was so determined that he looked at his own assets to start a farming program. He sold his own prized horse for only a

third of its value in order to start the program. Land was split into parcels and those parcels were offered to residents via an application system for farming. The struggling residents could come together, plant farms with crops of potatoes, beans and other vegetables, and help each other overcome this economic crisis and food shortage. Crops that first year were valued at over $14,000 and yielded over 65,000 bushels of potatoes. The next year, the crops' value doubled. The rest of the nation was watching Detroit and Pingree's plan. Pingree's potato patches became the 'Detroit Model,' which would be copied and implemented by major cities across the United States.[6]

Nationally, and honourably, Pingree became known as the 'Potato Patch King'.

But booms came again. And war. As a result, many of the urban gardens were abandoned, arguably because 'progress' meant farming was moved out of town, and city beautification schemes started to forbid livestock and some growing of crops. Dirty agricultural work was for the regions, not for the cities. Forty years later, hard times hit again, during the Great Depression, and again the leaders in Detroit encouraged so-called 'thrift gardens', which were a huge success. And then came another war.

Across the Atlantic, in Britain, the number of 'allotments' (personal gardens, separate from houses) skyrocketed at the outbreak of the First World War – increasing from about 440,000, to over 1.5 million three years later.[7]

These numbers shrank back gradually, to about 700,000 allotments just prior to World War II.[8]

However, the fresh outbreak of war in 1939 sparked a new, more concerted urban garden scheme. After Germany's successful attempts to stymie food imports, Britain started its government-sponsored 'Dig For Victory' campaign.

Necessity is a wonderful motivator. In 1942, only three years after the war started, food imports had halved – and Britain's allotment numbers had doubled again. Fully half the civilian population was involved in the nation's 'Garden Front'. This saw the conversion of about 10,000 square miles (2.6 million hectares) of land planted out to food.[9] Public parks, school playgrounds and factory courtyards were all transformed into allotments. The Royal Family sacrificed their rose beds for growing onions, and sporting ovals that weren't used as gardens were enlisted to graze sheep.

By 1943, there were 3.5 million allotments in Britain, producing over a million tonnes of vegetables.[10] Workers at Wolseley Motors in Birmingham made cloches out of scrap car windscreens for their workplace allotment. Even the lawns outside the Tower of London were used to grow food.

Like Britain's Dig For Victory efforts, Detroit (and much of the United States) also did amazingly well to feed itself during the Second World War. They did it under the 'Victory Gardens' banner. About 43 per cent of the nation's vegetables were grown in urban gardens during the war years,[11] which may come as a surprise to those who think of home gardening as incapable of feeding any great number of people.

Of course, garden waste wasn't wasted. It was upcycled by being fed to chooks and pigs. In Britain, a quarter of all eggs came from domestic hens during the war, and hundreds of thousands of people belonged to over 6900 'pig clubs'.[12]

Growing your own food isn't some passing fad. It's the way we used to do it. It's the way we've *had* to do it at various times in history.

The boom and bust of urban gardens again hit home after the Second World War, however. In the United States, it was claimed that progress was stymied by inefficient domestic growers, and it was in the national interest that people bought their food off big farms that used the modern technologies on offer, such as artificial fertiliser. Perceptions had changed, and suddenly efforts to grow your own food were seen as working against the economic recovery. The United Kingdom, however, was slower to shed the urban allotments as it suffered severe rationing for the best part of a decade after the war.

It wasn't until another crisis hit that Detroit once again championed officially sanctioned urban gardens. In the 1970s, when the automobile industry suffered setbacks from cheap imports, Detroit's mayor, Coleman Young, drew inspiration from the Potato King, transforming some 3000 vacant lots into gardens, under the clever moniker 'Farm-a-Lot'.[13] Nearly thirty years later, Farm-a-Lot was still getting about 2000 requests a year for help in setting up urban gardens.[14] Most of the food stayed within families or communities, and very little was sold.

Now, many may think of urban gardening, or backyard gardening, as a nice hobby, but of little value. A bit Kumbaya. However, much of the world's food isn't grown as a monoculture. Much of the world's food is grown on mixed, small farms. Some 60–70 per cent of the food grown on our planet is grown by smallholders[15] – even with a world population of close to 8 billion. Home gardens in the tropics can have over 100 species of plant and support nearly 1 billion people.[16] And if we all used the land close to us, the soil outside our back doors, to grow at least some of what we eat, the cumulative impact could be extraordinary. Remember how an area of land the size of the European Union is now used for cities? Some of this land could still be used to grow food.

Detroit's current urban farming model has different arms. Many tonnes of food have been given away. Some is sold. Much is used locally. The purpose of the gardens is to give people not only nutrition, and an excuse to be outdoors, but also a reason to socialise, to use their bodies in ways humans have evolved to use them. The fact that there have been over 300 varieties of fruit and vegetables on offer over the years, and that the Michigan Urban Farming Initiative has given away over 50,000 pounds (23 tonnes) of fruit and veg since 2012,[17] shows that the food side isn't just a nice feel-good thing.

It's not just some cursory and tokenistic moment for dreamers. It's real nutrition, going into real homes – often to those who are poorest and least able to buy fresh food.

Home gardens do way more than just provide parsley for your soup. There's strong evidence that where we live, our exposure to soil, affects our microbiome.

In Winfried Blum's 2019 article in *Microorganisms*, 'Does Soil Contribute to the Human Gut Microbiome?', the link is made clear. An urban existence removes us from soil, and from the genesis of much of our inner community. We simply don't get the bugs in and on us that we used to. This removal from soil has mirrored the upward trajectory in the incidence of simple autoimmune diseases, non-communicable diseases, and mental health disorders.

While the vast majority of our human microbes are different from those of soil, the two are far from disconnected. As Blum and co-authors conclude:

> Soil contributes to the human gut microbiome – it was essential in the evolution of the human gut microbiome and it is a major inoculant and provider of beneficial gut microorganisms. In particular, there are functional similarities between the soil rhizosphere and the human intestine. In recent decades, however, contact with soil has largely been reduced, which together with a modern lifestyle and nutrition has led to the depletion of the gut microbiome with adverse effects to human health.[18]

Other evidence backs this up. A study from the United Kingdom has shown poorer health outcomes for people with smaller domestic gardens.[19] Research from 2019 also suggests that the industrialised urban habitat is low in microbial biodiversity,[20] which means people aren't in contact with enough beneficial bacteria.

It's been thought a concept of 'microbiome re-wilding' from soil and plants could help lower the rates of lifestyle diseases that seem to plague urban settings.

Lots of people have grown food in an urban environment over the last hundred years of our exponentially increasing population. Following on from the thrift gardens during the Great Depression, and the Victory Gardens of both world wars, there have been concerted efforts to grow food in towns, with a spike in the last decade or two.

This is obviously not a new thing, and it's not going away. It's something we return to, time and again, when we want to use land – our soil – better. Urban gardening does meet council regulations head-on at times, but the truth is, if we love our soil, if we want to eat well, if we want to do more than just sustain ourselves, but also reinvigorate ourselves and rejuvenate our ecosystems, then growing food locally, very locally – even in cities – is a great option.

The thing about urban growing is that it cares for soil, and the ecosystems that soil can support. Sometimes a garden is purely for visceral pleasure; the shrubs and trees that make us happy. Sometimes it's groaning with fresh produce that can embellish any meal. Either way, underneath a well-tended garden the soil is humming with life, storing carbon, cleaning our water, freshening our air. And it can also nourish us physically, too.

Urban gardens – be they in backyards, on rooftops, in vacant lots, or in local allotments – are something to treasure and nurture. Caring for soil is a very personal contribution you can make to your well-being. It's a very tangible act, where what you do affects you mentally, physically, and perhaps even spiritually. It's intellectually stimulating, you're exposed to the mood-enhancing capacity of soil microbes, and the end result is food that is better not just for you, but for the world.

For too long, we've relinquished our responsibility for soil. We've entrusted the role of growing nutritious food to those who are paid to grow the biggest, brightest, longest-lasting food, not the most nutrient dense. But anyone can take part in the process of caring for soil, of growing food for the table – and that part can be as big or as little as your life, your energy, your circumstances permit.

Growing food isn't a fringe activity. It isn't unusual. It's not for everybody, but it benefits everybody nearby. You can take more care of your soil, feeding that microbial life and supporting a vibrant subterranean ecosystem, than any farmer can. You can also cycle nutrients straight from your kitchen to the soil, and back to the food you eat, using compost. You can nourish yourself, your family, your community, with an activity that we are physically capable of, that is good for our heads, our hearts, our backs, our immune systems.

Sure there are some constraints, such as water or access to land. If you're not sure whether your soil is contaminated or not, you may have to bring in some dirt to start with. But you think vacant lots in Detroit are chemical free? They simply use raised garden beds, which helps alleviate that problem, and also make gardening easier at the same time.

There are some general guidelines to remember about how to maintain complexity, like not disturbing your soil. Like covering soil, with living plants or mulch, whenever you can, to protect the ground and help it thrive. And you can do all kinds of tests that might show you what your soil might need, or do better with – but mostly, keep it simple, to keep soil life complex.

There is, however, one thing you'll do better with, regardless of where your garden is, or what you want to grow – and it's something many gardeners have known about for years. That is the power of compost.

CHAPTER 16

Compost, Compost and Compost

'There are three things I recommend above all else,' my partner Sadie says to the group touring her market garden. She's answering the single most-asked question we get on farm visits: how to improve your soil and your garden.

'These three things will make your garden more productive, more pleasurable, the soil more alive,' she enthuses. 'Just remember them in this order: compost, compost and compost.'

Sadie is speaking from experience. She now manages our market garden. She's the grower of 70 annual species in a kilometre of hand-tended, minimum-till garden beds, run on organic principles. Sadie, like so many other gardeners, reckons compost is key.

I know this, intellectually. And I've seen the results, practically. But despite compost's noble reputation among backyard gardeners, the allure long evaded me, and the science of it seemed impenetrable.

I've been building compost heaps for decades now, far longer than I've been growing food. Usually thrown together hastily, a compost pile seemed a better way to dispose of kitchen scraps than simply putting them in the bin. Often smelly, slimy, and the occasional refuge of rats, the big compost bin outside the block of flats in Melbourne where I was renting was a vile place; a site quickly visited and best avoided. A place to dump the contents of the kitchen bin and flee.

Compost, of course, is some kind of rotten organic matter, as I've always guessed. It's made of things that formerly lived, that are breaking down and becoming something else. But what is it, really?

An ancient forest, like the ones near me, the rainforests of a remnant Gondwanaland, smell musty. Of decomposition. But that smell is an earthy, ancient, primordial scent, hardly displeasing. In fact it's invigorating, and definitely not the smell of the compost bin outside that block of flats in Melbourne.

Good compost, of course, isn't rank. It isn't slimy. It, too, can smell sweet, like the scent of pleasant woodlands, not of stultifying decay. What is the difference? And what, exactly, is compost – and why is it so good for soil?

Well, the answer is as simple and as complex as you'd like it to be.

Compost is decayed organic matter, certainly. It's also a living system, with trillions of microbes. Most importantly, it contains the building blocks that soil life loves. All living things eventually die and decompose, their elements rearranged ready to be recycled.

Nearly 99 per cent of the mass of all living things on Earth is made from just six elements: carbon, nitrogen, oxygen, hydrogen, phosphorus and sulphur, in different forms.[1] And 99.9 per cent of all life is made from just 11 elements in the periodic table – the six above, plus calcium, potassium, sodium, chlorine and magnesium.[2]

These elements can be rearranged and recombined into new molecules, ready to be taken up by plants, or by the microbes that support the plants.

Whole ecosystems, the ones we enjoy now, are built on decay, on the earth's ability to reuse what's been living. Microbes are in the business of recrafting, repurposing and rejuvenating living and formerly living things. Plants are the land's genesis of energy, and everything else recycles that energy, with the loss of some dispersed heat.

Compost, when it's finished, is full of humus, which as we know is the dark bit of healthy soil (unless that dark matter is charcoal, as seen in the Dark Earths of the Amazon and Africa).

If you cut into your ground to see the profile, the darker bit near the surface is made by humus, which is essentially well-rotted organic matter – what was once called leaf mould. Humus contains really important compounds with names like humic acid and fulvic acid. It also contains carbonic acid, which is particularly useful for breaking down minerals in rock.

Humus in your compost heap comes from the same kind of composted materials that nature makes in forests, when leaves, animal matter, branches and all living and once-living things fall to the forest floor and are consumed by our wonderful worms and slaters, as well as microbial and arthropod life.

Humus is usually only found in the top 10–30 centimetres (4–12 inches) of the earth. In other words, humus is the bit that makes it topsoil, not dirt.

There can be millions of chemical compounds in humus,[3] all in very different states of decomposition or re-composition. Some of these have a very complex structure, chemically. Some polymers in compost – including glomalin, the high-carbon soil glue we met earlier – are tricky for the microbes to concoct. We don't entirely understand the mechanisms of their genesis, but there is coordination between all the microbes to make humus. They 'talk' to each other, knowing that if you make humus, plants will give you more sugar. Not immediately, but in the long run, because humus helps plants grow better.

No matter where on Earth it is made, humus, for some reason, ends up as pretty much the identical basic components. And the compounds in humus, such as humic acid, humins and fulvic acid, are like stimulants for underground life. They also do other fundamental things. Humic acid, for instance, is amphiphilic, meaning it is attracted to water and to fats, so it can help store them in the soil.

Humus in soil contributes to tilth, that wonderful textural ability of soil to hold air and not glug together, and at the same time not allow nutrients to flow through too quickly, to leach out. Humus in soil provides the perfect home and food for microbes. It also has a remarkably robust chemical composition that allows plant roots to function efficiently.

While compost is considered a great source of humus, the reality is we understand other things way more than we understand its complete role in soil. For years, humus could only be analysed by ruining it and breaking it down, its structure defying scientists' attempts to work out *how* it worked. For a long time, the easiest way to find organic matter like humus in soil was simply to incinerate it. The more heat soil gave off when burned, the more organic matter there was, because organic matter burns.

Any good gardener can tell you that compost works like fairy dust. But we still don't really understand why. That's despite years of research into humic acid, and its closely related humates (don't worry, you don't need to understand what they are – only soil scientists do). Humic acid was first identified in 1761, and isolated in 1786 – but 250 years later, the true chemical nature and entire chemical function of humus still eludes us.

When the talented organic chemist Justus von Liebig was pinpointing the role of NPK (nitrogen, phosphorus, potassium) as the limiting factors in plant growth in 1840, he also quashed the idea that humus was endowed with some kind of life force, called 'vitalism'. Instead of being some mystical carbon pump for plants, von Liebig showed plants get carbon from the air, not humus. Yet nearly 200 years later, we still don't really understand how humus works. For many of us, it might as well still be magic.

We just know humus is good for earth, and that earth will try to create it if left to its own devices.

To get a handle on how humus is made, let's think of trees, something more familiar. Trees are made of lignin – a tough polymer that makes wood particularly remarkable. Lignin gives wood enormous strength, as well as flexibility; think of the length and weight of a long branch, swaying in the breeze. Human engineers would struggle to make anything that is so simultaneously strong, pliable and beautiful. This tough polymer, lignin, is virtually indestructible by bacteria. It can only really be broken down by mechanical means (such as by insects chewing), or specialised fungi, called white-rot fungi. Even termites use fungi in their gut to digest wood. We know trees can make lignin, this resilient substance, out of the sugars they harness from the sun, and the micronutrients provided by the life in the soil. But we don't know how they do it.

The same is true for humus. We know it is, in part, polymerised carbohydrate – broken-down sugars that are transformed by our subterranean ecosystem with nitrogen. And we know humus is really stable compared to other components of soil, just as lignin is when compared to the cellulose in grass. It's harder to lose, in essence, and has a life force that we don't really comprehend. Two hundred years after von Liebig isolated NPK as essential for plant growth, the action of humus is still, scientifically speaking, a bit of alchemy.

Making compost is mimicking nature. If humus is the aim, and compost is the method, how do we make it, and why do we make it?

Well, we make it because we want the best soil we can get – and we also have things that would be considered waste, a problem to deal with, if they're not composted. This was probably even more true for home kitchens in times gone by, when food arrived in raw and not processed form, and we all lived closer to the land. It should still interest us, the recycling of waste, even if much of the production of food in developed nations has stripped the raw materials before it gets to us.

In fact, it should be easier to deal with large quantities at source, turning organic matter into humus, instead of deep-burying it or sending it to the tip.

The best way to make compost was properly examined only about a century ago. In the 1920s, a British botanist in India, Albert

Howard, wanted to work out how to improve soil to grow food using products already at hand, rather than buying in nitrogen or phosphates. It's the exact same problem many growers still face today. Can we reuse what we have and get better soil, without spending money unnecessarily?

Based at Indore, a research station in the biggest town in Madhya Pradesh, the central province of India, Howard did lots of composting experiments, harnessing the wisdom of the locals, and aided by a lot of indentured labour. Many think he was the first Westerner to record Vedic techniques of agriculture, a precursor to the modern organics movement. Scrupulously recording temperatures, materials and systems, Howard wrote detailed notes on how to make compost, for probably the first time.

His techniques were distilled into a booklet entitled *The Waste Products of Agriculture*, published in 1931,[4] the essence of which is still used as the basis for most types of compost today. Interestingly, right at the start of the book, Howard notes just how 'alive' healthy soil is with microbes, and how it's these tiny bugs that actually make compost.

Howard's practical work built on ideas from a book called *Farmers of Forty Centuries*, by U.S. Department of Agriculture scientist F. H. King, published in 1911.[5] King spent the first few years of last century travelling through parts of Asia, looking at how the locals could harvest crops year after year, century after century, from the same land while keeping it fertile.

King estimated that the Japanese were supporting three people per acre of agricultural land, compared to just a third of that number in Holland. How did they do this? They looked after soil.

On his travels through China, Japan and Korea, he found a multitude of ways in which the locals were composting, never wasting a thing from the land, returning all the ash, all the coals from fuel, to the earth. In fact, the government encouraged compost making where it wasn't otherwise commonplace.

Japan's Fukuoka prefecture, King observed, 'provided subsidies which permit the payment of $2.50 annually to those farmers who prepare and use on their land a compost heap covering twenty to forty square yards, in accordance with specified directions given'. He doesn't give the instructions, unfortunately – but he does have a recipe for barley fertiliser, which includes manure compost, rape seed cake and night soil (humanure). In some parts of Japan they handed out awards for the best compost heaps in the county!

Howard's work on compost was practical and observational. He honed how to successfully make compost, and his methods are still followed today. Howard was working in a region that had livestock, as well as plant residues (stems, stalks and other parts inedible to humans). He focused on getting the waste as broken down as possible, often mechanically, laying stalks and the like on roads where carts would drive over them. He then tested the layering of different types of waste, be it dried plant matter, or animal waste, or green plant mulch. And he was very aware of nutrient balance.

In composting, the two nutrients most at work are nitrogen and carbon. Together, these two are the basic chemical inputs required to get humus – and it's good if you can get the ratios in roughly the right amounts.

An overly nitrogenous compost is too rich, can easily turn anaerobic (airless, with no oxygen), and becomes slimy and smelly. That yucky smell is excess nitrogen being converted to gas, which isn't a good thing. This was often the compost I made when living in a flat.

Too much carbon and the compost can become too loose and mealy textured. It will degrade more slowly than is ideal, because the microbes need nitrogen in a certain proportion to work efficiently. Most microbial cells are about 50 per cent carbon, with the rest made up of acids and enzymes and protein, each of which could contain some nitrogen.

Ideally, a ratio of 30:1 carbon to nitrogen works best. That is, thirty times as much carbon as nitrogen. As the compost matures, this ratio declines to about 10–15:1, because the microbes, being living beings, convert the carbohydrates (sugars, containing carbon) into living energy, and dispersed heat, while also releasing carbon dioxide.

Things high in carbon are often considered brown – dead leaves, dried grass clippings. Things higher in nitrogen are green, such as freshly mulched leaves or grass, and food waste. Meat and manure are also exceptionally high in nitrogen, although the manure can vary a lot with the species. There are great resources on the web if you want more detailed information on this.

Howard's composting method, known nowadays as the Indore method, was very, very precise. But you don't have to be, because compost can sort itself out in the long run. That's what nature does, and it's the reason humus is essentially the same wherever on Earth it is made.

Remember, all organic matter is cycled. Composting is a way more efficient use of spent organic matter than deep-burying (which can release methane, and does nothing for soil fertility), or just letting it lie there, which releases more carbon and nitrogen back to the air, rather than storing them in soil as humus.

Compost can store about half the carbon that might otherwise be released into the air. And in the process, that living system magically creates humus – the soil superfood that makes our gardens better.

There's no waste in nature. If we can trap carbon for use in soil, we make the world a better place.

THE MAGIC OF COMPOSTING
Like soil – and like us, really – compost needs certain things to thrive. Food, air, water are the basics. The food is plant or animal waste. Air you get by regularly mixing and turning the compost. Your compost should also be damp, so wet it often. And it should warm up, as all those microbes get to work.

Most home compost is hard to manage if you add animal waste (meat scraps), unless you can keep rats and other vermin out. Using a bokashi bin – a Japanese-designed bio-digester – might be better in this case.

In terms of ratios, about twice as much 'brown' garden waste to 'green' is best. So, twice as many brown dried leaves as green lawn clippings, for instance. You don't have to be exact, but this is a good starting point. If you're unsure, having too much carbon (brown waste) is better for the smell, and for the environment, as it emits less nitrous oxide and methane. Think of brown waste as things that will burn cleanly – so, sawdust from old wood, compared to green, the tops of trees that have just been mulched. Or hay as brown, and freshly cut lawn clippings as green.

Common composting mistakes

Smelly, slimy compost: Too much nitrogen. You need to add more carbon in the form of dry leaves or dry grass clippings. You could add hay, but you might also be adding weed seeds if your compost doesn't get very hot. Early on, you could use sawdust or shredded paper, which will take longer to break down. A mix of different carbon sources is usually best.

Dry on top, smelly inside: Probably not enough air. Turn the compost regularly (weekly is good) to mix and aerate.

Dry right through: Not enough water, possibly? Wet it each week as you turn it. It could also contain too much carbon, but this is usually not the case for home gardens. Add green grass clippings, or more food waste.

Chunky compost: Possibly lots of big things from the garden that are hard to break down. If everything else is going well, you can make the compost and sieve these out, then return the big bits to your compost.

Cold compost: It's just working slowly. This will take a long time, so your compost may need more food, or more water, to jumpstart the microbial life. Hot compost can be ready in a matter of weeks, but cool compost can take months to mature.

When your compost is ready, it should have an attractive earthy smell, and crumble easily through your fingers. Everything small should have broken down into a very dark substance, which is essentially pure humus, ready to use on the garden. If you're not sure your compost is done, and are worried it might have dangerous bacteria (the kind you get in poo, rotting food and the like), you can still use it. Just shovel it around the base of fruit trees rather than your lettuces, so it doesn't get into your dinner.

CHAPTER 17

If it Quacks, is it a Duck?

Despite the seemingly mystical role of humus in soil, I'm a bit more hard-nosed than some when it comes to spirituality and farming. Whenever I hear statements like 'we plant our seeds after consulting the cosmos' or 'ecological harmony relies on the rhythms and cycles of the Earth, Sun, Moon, stars, and planets', I pray to a god I don't believe in to give me strength – or at least an open mind.

Growing food following the advice of people who have a different belief system, perhaps based on 'lines of energy' or 'pulses of life', is not really my thing. But the growing of food has long defined the human condition. A bad harvest meant a bad life. If the place you farmed or foraged was badly burned, or plagued with animals, your existence was pretty grim. If the soil was poor, you could be malnourished. We prayed to those who could give us more fertile land, more reliable crops, a better way of living.

Goddesses abounded, and we took particular note of those who cared for soil, and the ground. One of the few male deities was Tudi Gong, in Chinese folk religion, who was responsible for the harvest, amongst other things. Heqet was the Egyptian goddess of fertility, associated with the flooding of the Nile, and the germination of grain crops. She was often depicted as a frog sitting upon a lotus. Demeter was the Greek goddess of grains, agriculture and fertile soils.

Ceres was the Roman counterpart to Demeter, the goddess of agriculture, crops, fertile land and grain.

From the reliance on deities, to the ability of several cultures to improve soil fertility (as we saw in Chapter 13), soil has occupied much of our imagination.

We still imply that the Earth is a deity, and virtually always female. Mother Earth or Mother Nature, as we refer to her in English. Terra Madre in Italian. Ötüken in Turkish. Atabey in the pre-Columbian Americas. In Incan mythology, Pachamama is an earth mother goddess who watches over the sowing and harvests of crops. And there's Dewi Sri, the rice and earth fertility god of Bali and other parts of Asia. The fact that soil gifts us life, in the same way a woman can give birth to new human life, is a universal concept across cultures, across time.

But in the less esoteric, modern era, where science is the new god, and technology the new temple to worship in, these ideas seem quaint, odd, and perhaps a bit naïve.

Quaint, odd, and perhaps a bit naïve. That can certainly be said of Rudolph Steiner, the Austrian thinker, designer and philosopher who reframed education early last century. Steiner schools are known for being a bit 'out there' compared to traditional schools, places where strict time constraints and structured learning are often discarded in favour of refining a student's social, artistic and manual skills in a holistic fashion. They combine rhythmic physical exercise with imaginative learning, and emphasise whole foods. Rudolph Steiner had a strong spiritual bent, proselytising about a thing called Anthroposophy, which came through in his ideas on all things, education included.

What many may not know is that Steiner also rethought – or re-imagined might be a better way of putting it – the growing of food, and founded the Biodynamic movement.

In 1924, towards the end of his life, Steiner gave a series of eight short talks to farmers troubled by the rise of synthetic fertiliser. In these talks, he transferred his holistic approach from education to the land that gifts us life. He took a whole-of-farm approach, focusing heavily on soil health, and treating farms and gardens as defined organic structures, not just a series of separate entities, such as plant, animal, soil. Steiner, who by all accounts hadn't ever grown much in his life and certainly wasn't a farmer, expressed his ideas in broad, systems-based terms. The way he put it is that the energy within a farm should be retained where possible, so inputs should come from within the fence line, manure utilised, and animals fed on what is produced on site, not by what can be bought in. Doing this should imbue the whole place with more vigour, more resilience; healthier animals, more bountiful harvests.

The philosophy, in other words, was an antidote to some of what was about to happen to farming, globally, with the mass production of fertilisers and pesticides.

Steiner, however, also overlaid this philosophy with his own theological and mystical bent. A big picture thinker, he came up with some additional things farmers could do to achieve farm health, to meet his 'farm as an organism' goals. These are best summed up in Steiner's 'preparations', which are added to the soil where you grow food or graze animals. The purpose was to add a life force to the soil that – he suggested – wouldn't otherwise exist.

A couple of Steiner's preparations[1] are summarised here:

Preparation 500

Fill a cow horn with manure from a lactating cow, then bury it about 40 centimetres under the ground over the winter. The manure is said to bring calcium to the preparation. After digging up the horn, the contents are diluted with water. This is then sprayed on the soil four times a year, in the afternoon and during the descending phase of the Moon.

Preparation 502

The herbal plant yarrow is wrapped inside a stag's bladder and hung in the warm summer sun before being buried over winter and dug up in spring. The rotted yarrow (not the bladder) is used as compost.

These look like soil amendments, but are so heavily watered down that some might compare them to homeopathy. Interestingly, they do treat soil and plants as living systems. Steiner was speaking to people who weren't hard to convert – farmers who were often already Anthroposophists.

Steiner died in 1925, shortly after the lecture series. His work was taken up by his adherents in the 1930s, and became known as bio-dynamics. In the process of formalising a more structured approach, a whole bunch of practical aspects of what would become organics – which was also very interested in soil health – were incorporated into biodynamics.

Today, certified biodynamic farms – and there are many of them – differ from organics mostly in the addition of Steiner's preparations, with some focus on lunar calendar planting.

It's things like Preparation 500, the dung-filled cow horn, which many people talk about as the mystical attitude to soil. The woo-woo. It's easy to poke fun at them, and any peer-reviewed science to back them up is sadly lacking, even 100 years later. But in the era of greater understanding of soil microbiology, such things look a little less ridiculous than before.

For every advancement in knowledge, there's usually a charlatan taking advantage of our belief systems as well. But along with snake oil, there's been a whole bunch of people who work from observa-tions, from practical experience, who don't wait for the science to endorse what they do.

At the same time as Steiner was talking to farmers, a movement was well underway that started to consider soil in similarly life-giving forms, without the religious undertones. The organic movement, born in the United Kingdom, but cultivated using information from other countries with a rich farming tradition, such as India, was just about to kick off. While Albert Howard was dissecting ways to make compost in India, another Brit named Lady Eve Balfour used her not inconsiderable inherited wealth to explore the role of soil in plant health. In 1919, she bought a property, Haughley, that later became the first testing ground of organic agriculture alongside modern chemical-based farming, which became known as the Haughley Experiment. (The side-by-side comparison at Haughley wasn't over-blessed with scientific rigour – though it did show very positive results in terms of nutrients, crop yield, soil carbon and earthworm numbers.)

The seeds were planted (I was dying to use that cliché) for an organised movement to counter modern agriculture. Balfour credits Steiner with inspiring the idea of soil, and farms, as living organisms, rather than a series of separate entities. Balfour's book *The Living Soil*, published in 1943,[2] exposed many to the idea that soil was indeed alive (I guess the clue's in the book's name). Balfour followed that success in May 1946 by founding Britain's The Soil Association, widely regarded as one of the first soil health–focused groups in the world. She became its inaugural, and long-serving, president.

While most of the agricultural industry in the modern world was heading towards inorganic fertiliser use and hybrid plants for higher yields, Balfour and others kept the idea of soil life in mind. Out of this was born the modern organic movement. While organics, like biodynamics, is focused on whole systems, not just soil, the principle of all organic production is grounded in soil health.

Balfour's impact on the soil movement in the United Kingdom was vast, but Britain wasn't the sole beneficiary. The tendrils of the movement quickly spread around the world. Within three months of the inauguration of Britain's Soil Association, on the other side of the globe in Australia, hop-grower and local soil enthusiast Henry Shoobridge set up the Living Soil Association of Tasmania. The Tasmanian outfit was the first international affiliate of the British group, and the fifth such organisation in the world.

If Balfour was inspired by Steiner, then her visit to my home state of Tasmania in the late 1950s kicked off even more interest in soil health and soil-focused growing. In Tasmania, there has long been interest in quite varied agriculture, and the island state is noted for its devoted home gardeners. This in turn has fostered a strong organic growers group. While the Living Soil Association of Tasmania didn't really outlive Henry Shoobridge and the 1960s, it was superseded by the Organic Gardening and Farming Society of Tasmania, in 1972. It was groups like this, which also popped up in other parts of Australia, that helped cultivate the next generation, such as Peter Cundall, Australia's organic gardening elder statesman. For a generation of Aussies, Tasmanian-based Cundall was the face of national broadcaster ABC's *Gardening Australia* weekly television show. Tassie is also home to Steve Solomon, whose nutrient-focused approach to soil, and his gardener manifesto *Growing Vegetables South of Australia*, are informed by all that came before. From Steiner to Balfour, to Shoobridge to Cundall and Solomon – it's virtually a straight line, despite the loss of the theological bent.

Tasmania is also, and probably not by complete coincidence, the birthplace of permaculture. Bill Mollison and David Holmgren, permaculture's founders, met in Tassie. Mollison, a professor at the University of Tasmania during the 1970s, was born in the north of the state, and would have lived with some of the concepts being explored by a plethora of local organic growers. Holmgren, a postgraduate design student at the time, probably spent many a cold Tassie night in the early 1970s pondering growing systems with Mollison, and ideas such as the energy flows wasted in conventional agriculture, the futility of annual crops, the wonders of soil, and if there was a better way to manage things.

The term 'permaculture' is an amalgam of the words 'permanent agriculture'. It has its own belief systems, which are more flexible than Steiner's – but the idea of living soil, and that soil begets life, continues unabated.

From our original deities, the goddesses of the Earth, we've come a long way – in some respects.

Starting about a century ago, Steiner's quasi-religious 'farm as organism' concept has birthed a range of groups all looking at soil, despite their differences in some core beliefs. Biodynamics is a distillation of Steiner's ideas, but made more practical. Organics follows most of the soil health ideas of biodynamics, without the prescribed preparations. Permaculture uses soil as just one idea in a framework of ideas, but it wouldn't be possible without the fundamentals laid down by generations of organic growers.

At all stages there have been visionaries. Sometimes they're quacks. Sometimes they're abrasive. Sometimes they get it wrong. Occasionally they focus more on faith than actual results, and certainly many of the arms of the soil health movement are more observational than simply reliant on science; on what they would argue is *reductive* science. But all the time, from Steiner to Holmgren, they've circled around to focus on soil as the original sustenance. In this way, they've differed from more industrial farming systems. For a while in the mainstream, soil has been seen as physical and chemical, but the biological – the living part – has often been neglected.

That, however, is on the cusp of changing. Seeing soil as *living* never went away; it just got sidelined for a while.

The good news is that soil is finally now getting the attention it deserves.

CHAPTER 18

We Are All, Temporarily, Not Soil

A woman's body is slumped over a log. A shallow grave with a man's body in it lies nearby. Not far from both, a tarp covers a bloated corpse left to rot in the sun. It sounds awful, if not downright barbaric. But this isn't a grim crime scene. Rather, it's a research station where human cadavers are the focus.

Over 1700 volunteers have had their corpses laid out in various circumstances at the University of Tennessee's Forensic Anthropology Center, ostensibly to rot. The aim is to see what happens, in different scenarios, over different time frames, to the human body, to help solve crimes and answer archaeological riddles.

Why am I telling you this? Because, quite simply, we are all, for a certain while, not soil, and the best research we have on decomposition comes from body farms. The atoms that make up our bodies were once, and will be again, the land that grows things. Everything that we are now has been cycled through soil, whether we're Buddhist or atheist, cremated or buried, Sudanese or Scottish. All that carbon, oxygen, nitrogen, hydrogen, calcium – it all came from somewhere, and it has all been in this cyclic loop for millions upon millions of years.

All non-photosynthesising creatures, be they single-celled, or slugs in the garden, or complex, milk-producing mammals like the great apes, are reliant on something that has once lived. What we eat depends a lot on our place in the food chain, because what we eat is made of all that came before. And when we die, we become something else's food as we return to soil.

It's not like this is a mystery. Yes, we know about pushing up the daisies, ashes to ashes and dust to dust, but having a dedicated academic study facility, a 'body farm', seems macabre. At Tennessee, and similar research institutes, they are trying to work out how we rot, to demystify things for police, coroners and the relatives of those whose remains are found, but whose stories may not be known.

I'm interested in this because dead things, including us, are . terrifically good for soil. A side benefit of research into the way bodies break down for forensics is that we can also see the positive impact on soil fertility.

Now, this won't be a surprise when you think about it. Most people have heard that blood and bone, a common soil amendment, is great for fertility in the garden or on crops. We know that animals must die, in nature, and be returned to the earth – first by scavengers, maybe, or carrion eaters and insects, and then by bacteria and other microbes. The impacts of that rot are fascinating.

Work in Tennessee and elsewhere has confirmed that when a body is laid on the ground and decomposes, the soil does become much more fertile. But only in a certain area. Bodies, humans included, form what is termed a 'cadaver decomposition island' – an area that takes up all the benefits of nutrient-dense flesh and organs, and converts it to soil fertility. This decomposition island is about the shape of the original body, with what is disconcertingly called the 'maggot mass' around it. Imagine the archetypal crime scene, with a body outline inscribed on the ground as the detectives peruse the location. The decomposition island, the bit that has way more going on in the soil, is in this shape, but a bit fuzzier at the edges. The vast majority of the fertility doesn't spread out from there any time soon.

According to the scarily but appropriately named Corpse Project – which asks how we can lay our bodies to rest so they help the living and the Earth – cadaver decomposition islands, or CDIs, are:

associated with increased soil microbial biomass, microbial activity ... and nematode abundance. CDIs release energy and nutrients into the wider ecosystem and receive materials such as dead insects and faecal matter and feathers from scavengers. CDIs are a specialised habitat for a number of flies, beetles and pioneer vegetation, which increases biodiversity. Increased soil carbon, nutrients and pH is present in a CDI during advanced decay. Bison [corpses] can affect the structure of plant communities for at least five years.[1]

What that means is that dead things are really good for soil, and ecosystems more broadly, even if much of that benefit is confined to a very narrow area around where the body rotted into the ground.

Cremated remains, The Corpse Project notes, are not very good as a soil amendment, with the main constituent, calcium phosphate, not available to plants in that form, and all the carbon and nitrogen is burned off during the process of burning the body.

The cadaver decomposition island is an interesting if gruesome concept. It shows how soil fertility, and soil microbial life, can vary a lot within a very small distance.

Cattle farmers can see this same impact all over their paddocks, where high fertility is conferred on a limited area under poo. A big cow pat can cause a flurry of green growth around it, and eventually through it, over the subsequent couple of years, in contrast to other patches in the same field.

Why does this matter? Because fertility is relatively static – but living soil, especially when it contains lots of fungi, can balance out the discrepancies, as the fungal hyphae seek out nutrients far from the plants.

With a population of nearly 8 billion people currently alive, and who will all die, there's a lot of soil fertility being squandered that we could use. It might seem disrespectful to consider composting human bodies … but is it really? Already, graves in Greece are emptied so new bodies can be put in, only a year or two after they're interred,[2] because they need the graves for the newly dead. Using cadavers sounds morbid. Nobody likes to think of nature's processes breaking down our own bodies while we're still alive. It makes me squirm as I write this, and could well make others agitated or unimpressed. But wasting fertility may not be an option if we want to keep so many people living in the world.

What using human bodies for fertility might look like is still a big unknown. It would involve shallow burial, in rapidly decomposable caskets or shrouds. It has to recognise the dignity of the once living, and respect the wishes of those left behind. But it's something to consider if we want to give back to the soil that gives so much to us.

We, of course, are just one of the animals that exist on growing lands, despite our ability to dominate landscapes and change environments. Humans are only 0.01 per cent of the biomass of all life on Earth.[3]

Ecosystems, even the ones we have globally now, after fairly large alterations by humans, are built on various trophic, or feeding, levels. These levels include animals. Soil is built in this majestic nutrient cycling between the atmosphere, plants, the invisible life below and above ground, the small things like arthropods and worms and insects, and the bigger things, grazing animals, predators and birds. Take any of these out of the ecosystem and it can struggle.

Cropland, and in fact most agricultural land (including grazing land), is not a natural ecosystem. But, as we've seen in Chapter 13, very few landscapes are not managed, in some way, by humans. The Amazon was a human-managed system, when ancient communities first discovered a way to make *terra preta*. Australia was well managed by humans for 60,000 years or so, under a strict regime of forage and rest, firestick farming and recovery by Aboriginal people. Britain is and has been modified for thousands of years, as has the United States by Native Americans, and so too has every long-inhabited land.

It's only recently, however, that we've removed animals from some of our farming systems, which has implications for nutrient cycling.

It's widely recognised that all the ungulates, those animals with a hoof – so, ruminants and other grazing animals such as horses, ibex, giraffes and zebra – have a substantial impact on soil nutrients.

Swedish researchers, for instance, looked at the importance of reindeer (their local grazing animal) on soil health. According to their 2016 paper from Umeå University: 'Reindeer do not only influence soil nutrient availability, but also by which mechanisms plants acquire their nutrients.'[4] They note that herbivore dung and urine can drive the diversity of plants, as well as how much nitrogen is stored in the soil – but only where you have a certain density of grazing animals.

When bison, more commonly called buffalo, roamed the United States, it was conservatively estimated that 30 million of the beasts chomped the rangelands[5] and pooed and died there. Some say 60 million.[6] As they migrated over vast swathes of North America – over 60 per cent of the landmass – they spread manure, and their fertility-laden dead bodies, across the landscape. Thanks to human-driven slaughter, numbers dropped to only 325 wild bison in the late 1800s. These days, while the bison population has rebounded to about 500,000 animals, they only roam over about 1 per cent of their former range[7] – so the land has lost a massive amount of nutrient cycling to and from its soils.

Science has been showing how soil microbes, and soil carbon, can be enhanced using grazing animals. The simple act of grazing, if managed well, can trigger plant growth at the right times and alter landscapes. It's all about enhancing nutrient cycles, and keeping as much of the carbon as we can in the earth, where we want it.

Grasses are meant to be grazed. They put down most of their growth below ground, in their roots. Grasses have different exudates – the chemical compounds that are fed to the microbes beneath the soil – than crops or trees.[8] When a plant is damaged, which it is designed for, it sends chemical signals, and food, to its roots, to aid in repair. Nobody's 100 per cent sure why, but the exudates you get from mowing grass aren't as plentiful as those you get from grazing. Scientists think that because a grazing animal pulls at the pasture a little, this tears off microscopic roots more than cutting it does. The grass then sends out something akin to an emergency signal to its underground community, releasing more – and more variable – exudates (microbe food), and the microbes respond accordingly. Effectively, grazing awakens the underground ecosystem.

Grazing animals foster grass growth by tearing off grass stems when feeding on pasture. Some also have chemicals in their saliva that act to trigger more vigorous plant growth, too. Grasslands and grazing animals are meant to go together, ecologically: what comes out the back end of an ungulate, the manure, also comes ready armed with microbes that are good for soil health.

As animals have been removed from landscapes, a lot of nutrient cycling has diminished. Not all has been lost, however. Even today, it's estimated that globally seabirds move up to 100,000 tonnes of phosphorus from sea to land annually, a natural consequence of their feeding and defecating.[9] In scientific circles, animals that do this are classified as 'ecosystem engineers'.

Bats, rodents and insects are all ecosystem engineers because they all move nutrients around. In fact, studies show that rodents can be as beneficial for grassland nitrogen availability as large grazing animals, such as horses or cattle. Many of those little animals have the 'pepper shaker' effect, where small amounts of faeces are spread over a large area, as the animals move around. Localised 'hot spots' can emerge where the animals roost or sleep, but the pepper shaker effect means it can be hard to measure the amount of benefit they confer over large spaces. That's particularly true of creatures such as small temperate-climate bats. Big animals are better researched.

What we do know is that the pepper shaker effect does increase organic nitrogen availability, a good thing for soil, and this is more likely to happen when the farm is a functioning ecosystem, with trees and wildlife corridors, as well as managed paddocks.

Since humans first walked the Earth, there have been massive shifts in the cycling of nutrients.

In a brilliant 2015 study on nutrient cycling by large herbivores, Christopher Doughty and his team estimated that the capacity of land animals to move nutrients away from concentration patches has decreased to about 8 per cent of what it would have been prior to the extinction of megafauna.[10] The vast majority of this happened from about 13,000 years ago – just as agriculture was taking off.

In other words, just by removing large grazing animals out of the equation, we've stripped 92 per cent of the nutrient cycling that enabled ecosystems to function naturally.

Grazing animals and land animals are just one source of nutrients. Some estimate that 85 per cent of terrestrial and freshwater aquatic biomass is fuelled by marine-derived nutrients, moved there by birds or fish.[11]

How does this work? Salmon are the classic case study, because they are a migratory fish, and there's been lots of research on them in North America over the years. As Scott Gende and others wrote quite succinctly in an article in *BioScience* in 2002:

Bears foraging at streams in British Columbia move 58 per cent to 90 per cent of all salmon biomass to land, sometimes

hundreds of meters from the stream, and further distribute
the minerals and nutrients in the form of urine and feces as
they move throughout the riparian and upland forests. Stream
insects feeding on salmon carcasses often have aerial adult
phases, during which they can fly far from the natal streams.
Avian scavengers [birds] remove chunks of salmon tissue and
carry them onto land and also leave their excretory products
across the landscape.[12]

Over half the weight of salmon – in some cases nearly all the salmon –
was brought up onto land by bears. It is also moved miles further
away from the source by birds and insects, to cycle in soil even higher
up the hills.

Nutrients are always on the move, when you have animals.

In Alaska, grasslands have been converted to tundra when predators have been introduced, resulting in reduced seabird colony
numbers. Get rid of the birds, and the plants change, because of a
depletion in the available soil nutrients when there's no bird poo.

Much of the modern research on nutrient cycling is on non-
agricultural land – into things like seabirds roosting on islands – but
overall, the implications are huge. If we remove animals completely
from ecosystems, which is happening as we create more and more
monocultures, this decline in nutrient cycling is likely to worsen.

Yes, the last ice age did give us more minerals to work with in
some areas. Yes, we have worked out the rudimentary chemical needs
of plants. But as we transform landscapes, or overfish, or pollute
rivers, or cut down forests, there can be flow-on effects that take over
a generation to see.

The signs are already there of the damage our increased hunger
for land, and for fast results, has done to once quite mobile nutrients
in the system.

We've been pretty good at reducing most animals, both in their
numbers and in their movements – including small creatures many
of us never really think about. Slugs, the bane of many gardeners,
use up to 100,000 tiny teeth on their tongue-like radula to turn leaf

matter – particularly dead leaf matter, which they quite often prefer[13] – into nitrogen-rich, mineral-heavy compounds that are great for soil health and plant nutrition. Other small creatures, particularly insects, break down leaf litter too. The loss of insects, through wholesale insecticide use, means 'frass', or insect poo, is lost from the cycle. As one research paper puts it, 'Frass deposition to soil is an important pathway by which herbivorous insects impact decomposition and soil nutrient availability.'[14]

Imagine a stand of stubble – the stalks of barley, say, after the crop has been removed. Usually the plant has died, or often it's been 'desiccated' with a herbicide such as glyphosate to dry it off, so it's no longer living. This stubble will decompose, but slowly. It will be blown off as atmospheric nitrogen and carbon dioxide, feeding little, storing no carbon in the soil, returning to the atmosphere. Oftentimes it is burned instead, which puts the carbon and nitrogen into the air even quicker. But insects, despised by most growers, can decompose the stubble in a more helpful way. Slugs break down dead leaf matter into smaller parts. Mites also break down plants that slugs may not favour, again breaking them down into tinier and tinier pieces, so that fungi can more readily transform them. Worms digest above-ground leaf litter and turn it into soil gold, feeding and transporting microbes, and storing carbon and aerating soil as they go.

The more things that are broken down into smaller parts, and are kept in the nutrient cycle, the more soil health we end up with.

Considering that the nutrient cycling between bacteria, fungi, archaea, nematodes and amoeba are poorly understood, and little known even in a small area, it would hardly be surprising if large landscape-scale understanding of nutrient cycling defied science for some time.

We would, however, have a better understanding of this incredibly complex story if Gary Polis was still alive. Polis, an ecosystem scientist, was fascinated by the ability of nutrients to cycle far from their source, from oceans to deserts, and was particularly talented in the field. Unfortunately he perished when a boat in which he was travelling with four other scientists capsized in bad weather in the Sea of Cortez in March 2000. Polis and the others were on their way to study islands in the Gulf of California, to research the seabirds that moved nutrients onshore and inland. In the book he co-edited, published posthumously, called *Food Webs at the Landscape Level*,[15] Polis explored nutrient cycling by aquatic insects onto land, by mice and deer into pasture, the effects of agricultural subsidies on snow geese and the resultant diminution of their natural habitat.

His research, and that of others in the book, shows everything we do to soil, we do to the world as a whole, and vice versa. Treat the world with respect, and it will gift us life, in all its forms. Muck about with soil, destroy it, or ignore its current plight, and the consequences are larger, and more far-reaching, than we know or can predict.

'I feel bad,' says Juan, our cook who lives off the electrical grid on a bush block not far from our farm. 'I don't like it,' he complains after spending a whole day away from home. 'When I have to do a poo at someone else's house, I think about how I'm losing the nutrients that should be on my place.'

Juan has a composting toilet at home. And an unusual attitude to how best to maintain fertility on his land. But he's not alone. Using poo, be it human or other animal waste, is a really big thing that most of us don't do. Not using doodoo is what we tend to do.

In previous chapters we've seen how animal waste is a great thing for soil health. I've talked about how wild and domesticated animals cycle nutrients from sea to land, and from fertile to less fertile areas, albeit in ways dwarfed by the now extinct megafauna. Many people have long seen the value in what comes out the back end of not just livestock, but also ourselves. For much of human history, people have used turds and wee to turn poor-quality soil into something better to grow their crops. *Terra preta*, black earth, was probably built partly on the judicious use of human poo. Humanure, it's called these days, and for most of us a ghastly topic it is. Much easier to just flush it away and never see it again.

But all that sewage is nutrient-dense waste. It's what soil has gifted us, in very condensed form – all those minerals and micronutrients pass through us, apart from the relatively small amount we absorb permanently. In theory, this waste could be composted and used as a soil amendment. Yes, there are problems with the antibiotics that are also flushed through our systems, and flushed down the loo. And yes, wastewater can also contain illicit drugs. (In Australia, the authorities test wastewater to get an indication of how many in the population are using drugs.) Sewage can contain heavy metals from industry as well, because the wastewater system is used as a garbage can by many.

But, really, using human poo as a fertiliser isn't all that unusual. Until about a century ago, most people in Western nations didn't have flushing toilets. Most people, globally, still don't. Up until the late 1800s, human poo was as common in the garden as any other kind of manure.[16] Talk about humanure these days, though, in any but the most eccentric circles of hardcore preppers or gardeners, and you'll be looked at like you're shit on a shoe.

Not everybody wants to poo in a bucket. The thing Juan wants to save is not for the faint of heart, but it is something we need to think more carefully about when soil health is decreasing, and our population is increasing. Humanure can feed the soil as well as the other way around. So to speak.

Not all poo is wasted. About half of America's sewage effluent is used on farmland. In Sydney, currently Australia's biggest city, effluent is dumped, after only minimal processing, into the ocean. Every day, 500 million litres of strained sewage is pumped out of pipes off Sydney's coastline.[17] That's about eight Olympic-sized swimming pools an hour. The effluent contains important nutrients removed from soil and dumped far from the source. It's not quite the days of the Bondi 'brown trout', the local euphemism for raw sewage that used to seep onto the nation's most famous beach in the 1980s, before they extended the outfall pipes further offshore. Today, the same stuff is just filtered a tad more and pumped further out. It's still far from ideal – 500 million litres a day from ideal.

Canberra, Australia's capital and the city I grew up in, does better, as do most inland towns. They have to. They can't rely on the ocean to clean up their mess. In Canberra, they tertiary-process the sewage, meaning it's as clean or cleaner than the local rivers, then pump the water back onto parklands and sporting fields. They also irrigate about 100 hectares of grape vines with it.[18] The 'solids' are burned (releasing the carbon, oxygen, hydrogen and nitrogen into the atmosphere), and the resulting ash (mostly minerals by this stage) is spread on farmland. It's inefficient in some ways, but at least the waste is used. Apparently, the use of the waste without incineration is considered socially unacceptable, and is a non-starter because of perceived risk from the bacteria.

Nobody is suggesting we put human poo, or any animal poo, in its raw form on our vegetables. The *E. coli* and other bacteria in it can pose a real risk to human health. But composted into proper humanure, it could be a boon to soil fertility.

Talking about death and poo are relative taboos in our society. We don't like thinking and talking about them. Flush it away, bury it, hide it and don't think about it too much.

But we know how dead animals can make soil fertile, from the cadaver decomposition islands of the body farms. We know that human excrement, like all animal excrement, is just one process in the cycle from soil, to plant, to animal, to soil again. With topsoil loss and a growing population, we may have to rethink how we use these resources in a way that will benefit the earth that feeds us.

CHAPTER 19

Keep Them Dawgies Movin'

Allan Savory, ecologist, once shot 40,000 elephants.[1] Well, he didn't shoot them all personally, but he was responsible. In his role managing rangelands in southern Africa, Savory ordered the mass slaughter of the world's largest land mammal, an unusual strategy for one interested in restoring nature to balance.

How did a one-time Rhodesian soldier turned ecologist come to the point where he thought it a good idea to shoot elephants? It's the same story that has led to a revolution in the way we see ruminants, and grazing.

In the 1950s, Savory was tasked with identifying areas in Africa that had good conservation values: places earmarked to become national parks. The areas were protected from humans, mostly – and in particular humans with guns who would shoot big game. But within a relatively short period, the areas were becoming denuded of plants, and rapidly becoming deserts. Savory blamed the elephants for overgrazing, and decided the only way to protect the savannah was to kill them. He did so with the blessing of the best scientists on the ground at the time.

Savory didn't just shoot 40,000 elephants and forget about it. First of all, the land where the elephants had been continued to decline in quality. Then, much later, Savory noticed that some areas of soil on a sheep ranch near him had improved, whereas other patches hadn't. Where the animals had bunched together during a storm, then been moved on, the land had improved, with more diversity of plant species, more leaf litter on the ground. Better soil, in other words.

As Savory told the United States' National Public Radio, NPR, 'The land had suddenly improved. And that clicked in my mind. And I realised, oh, my goodness, we can use livestock perhaps to mimic nature and mimic the wildlife herds. So let me work out how to do that.'[2]

The problem with the elephants on reserves, Savory theorised, was that they could no longer roam as they once did. They were stuck in the one area, free from human hunters, but also free to continually graze the land that fed them. So Savory set about trying to work out how livestock can not only denude and damage grasslands, but also do the reverse; restore them. He saw how wild ruminants bunched together, to avoid predators. They moved from one place to the next in mobs, so only grazed each area for a very short period in high numbers. As they did so, they were pooing and trampling grasses, and they didn't return to a grazed area for a vastly longer period than it had been grazed.

The theory goes like this: grass grows rapidly if not grazed too short. It puts most of its carbon into the ground in this period, before it gets leggy and slows down. If you graze it just as it starts to decrease in growth, you get a pulse of food for the underground life from root exudates. The animals' feet scuff the ground just a bit, they trample any long or stagnant grass and poo on it, creating a mini compost heap against the earth. All the soil's big things – worms and mites and arthropods – along with the little things, like fungi and bacteria, start to digest this mat and turn it into soil, burying it as they go. After a day or two of grazing, the animals then leave the grass to rest and build energy again.

These ideas, mimicking the nature of nomadic grazing herds (and in part the theory of composting), have been adopted in many areas. Lots of pastoralists, graziers and shepherds now move their herds or flocks, trying to emulate the great migrations of herbivores from yesteryear.

Allan Savory's name will be forever tied to the elephant carnage. His name is also now tied to a concept that has been sweeping the world's grazing community of late – rotational grazing, the movement of livestock in the hope that it will lead to better environmental outcomes.

Savory is such a powerful advocate of rotational grazing, and the power of farming to do good, that a TED Talk he gave in 2013, titled 'How to fight desertification and reverse climate change',[3] has been watched online over 7 million times.

Vast swathes of the Earth, about a third of all our agricultural land, is in danger of becoming desert because of overgrazing. This new rotational grazing method, actually invented by a little-known French farmer and researcher, André Voisin, but popularised by Savory, promises better results.

Because it's done for short periods, grazing large numbers of animals in a small space, it's also known as mob grazing, or cell grazing. Different names, but the basic theory is the same. Instead of what is known as 'set stocking', where a certain number of animals are plonked in a paddock for weeks or months at a time, the animals are moved daily, or every few days, to mimic wild herds. As a result, the grasses aren't eaten too close to the ground, and the pasture is then rested for weeks, months or even longer before being grazed again. In the rest period, grasses gain vigour and can re-establish themselves – thanks, in part, to the animal manure that was deposited there.

In conventional set stocking, however, animals can graze their favourite plants over and over, weakening the roots. The root tips die off, and can't penetrate as deep or travel as far as they did, so when dry times come, the plant has less access to water and nutrients. The livestock's poo is more spread out and not scuffed around as much, meaning, with cattle at least, that it will take longer to break down (especially in drier areas). And in set stocking, the theory goes, the grass never really gets going again, never able to fully ramp up its photosynthesis and punch carbon down into the soil, because it's constantly being cut short, never able to make as much sugar out of thin air.

Savory's main body of study is now enshrined more broadly in trying to work out what the farmer, the farmed, and the environment need in any particular area, and managing for all of those. It can, and does, include moving larger mobs of grazing animals from place to place. But that's only part of the larger whole of what he calls Holistic Management (his capitals), a system that incorporates Holistic Planned Grazing.

As I write, there are 13 million hectares of grazing land under Savory's Holistic Management,[4] with sites on every continent that has grass. He's either right, or very charismatic – or the conventional system is so faulty that pastoralists are desperate for a soil fix. Or perhaps a combination of all three.

Savory's claims are pretty big. He says that using his systems, we can re-green deserts. That we can repair damage to fragile arid landscapes, and improve the fertility of some of the most marginal land on the planet. Savory claims we can reverse climate change by having *more* ruminants on Earth, not fewer.[5] The theory is that if we scale up those sheep (and other ruminants) and restore grassland, we can sequester carbon, build soils and re-green deserts. In other words, heal the world.

Some in the scientific and grazing community are highly critical of Savory. Rangeland scientists from the United States even wrote a rebuttal of his famed TED Talk, noting his failure to demonstrate scientific proof.[6] They also point to his use of contested photographs to illustrate his point, though their tone does sound bitter and injured at times, rather than just reasoned.

Evidence doesn't always support rotational grazing, at least in simplistic terms. Perhaps it comes down to the actual detail of the study, and of each particular farm. It's hard enough to test one piece of land, let alone multiple sites, over multiple years, for a single factor (say, rotational grazing), when other factors (land use change, seasonal variation, changes in fencing, and so on), can also come into play.

It certainly seems to come down to rainfall. Most of the science showing that livestock can actually improve carbon in soil is based on higher rainfall areas than Savory tends to espouse can be brought back from desert.[7] Soil reaches a tipping point. Once it loses too much carbon, or too much life, or too much of any particular part of its structure, it can be much, much harder to restore.

In Australia, where rainfall is traditionally very poor, soils are weak and old, and soil carbon is generally getting lower by the year, there is great work being done in this field. So too in Africa and parts of the United States. If you look up the studies, rotational grazing has been shown to:

a) not work

b) possibly work

c) work really well

d) work with particular rest times.

In other words, the jury's still out. But talk to farmers who practise it, and there's no doubt. They walk the land, they can hear the bird calls (which, by the way, they hear more of; there are studies showing rotational grazing can increase the presence of song birds), and know that what they are doing is changing the land. If I had to shed all my years of farming and pretend I had no preconceived ideas about it, the science isn't indisputable. But it is compelling. And the research is getting more nuanced and more convincing year by year. The worst results show little carbon storage, and few environmental benefits compared to conventional grazing. The best show regenerated carbon levels and healed landscapes. The science is pretty much settled: well-managed grazing can store carbon in soil, even if individual results vary.

Long-term studies on the impacts of cattle and sheep are few and far between. But studies on other ruminants can be enlightening. Research looking into reindeer shows that they, like all ruminants, can drive vegetative change. A 50-year study found reindeer can alter a landscape of moss and dwarfing shrubs to forbs (flowering meadow herbs) and grasses, while storing nitrogen in the soil.[8]

Every year sees more convincing studies – from the United States, from Africa, from Australia – showing that, yes, rotational grazing certainly can store carbon in the soil more than many other styles of grazing. On top of research into carbon storage, there are other studies of a more general nature. Rotating livestock in a mob grazing pattern has been shown to even out the animals' diet. Instead of eating only the most palatable plants (and often more tender, more vulnerable plants), some research on the plains of Africa suggests that rotating livestock forces them to eat the more tannic plants, too.[9] This takes grazing pressure off the sweeter plants – and, in a nice piece of synergy, more tannic plants produce less methane in an animal's gut. A varied diet also gives the meat more flavour and nutritional nuance.

Rotational grazing can change plant species in paddocks in positive ways. It can benefit biodiversity on the farm more generally, too. Results depend on individual differences in geology, geography, rainfall and management.

The evidence for guaranteed success is dependent on other factors besides simply moving animals around. As one scientific paper from 2016 puts it, 'very few studies take an ecosystem approach and none have considered rangeland food webs'.[10] The research – proper, peer-reviewed scientific studies – is hard to conduct. Good farm managers are doing many things all at once, not just moving cows. It's a far cry from a lab, a Bunsen burner and a woman in a white coat doing titrations on chemicals, the way 'pure', simple science is conducted.

Adaptive farmers see their land as a whole, balancing the needs of all who use it – humans, livestock *and* crops. They plant trees. They rest paddocks. They fence off waterways and create leaky weirs to store water in the landscape longer. They are also very cognisant of soil. Often, and heroically, putting soil first.

Anyone who grows food is most impressed not by seeing a peer-reviewed paper, but by seeing an actual peer, another grower, in action. Anybody interested in the topic might just talk to others who have adopted a whole-of-farm approach – often regenerative farms, such as the multi-species White Oak Pastures in Georgia, Charles Massy's sheep farm Severn Park in New South Wales, the Haggertys' grain-growing properties in Western Australia, and countless others.

If you want to see growing land as ecosystems, not just as pages from a child's storybook or photos from a fertiliser brochure, take up the opportunity to visit such farms.

Savory is in esteemed company as an outsider. A lot of people who first suggest something that doesn't fit the reductive scientific model are considered cantankerous. Prickly Pioneers, one of my friends calls them. Rudolph Steiner could probably be considered one, ridiculed outside of biodynamics for his cow horns filled with manure. Peter Andrews, a man who re-imagined water flows through the Australian landscape,[11] is another, squabbling with water scientists and bureaucrats while showing actual results on the ground. Bruce Pascoe, Indigenous historian and now grain grower, is another, when he demonstrated quite clearly that Australian Aboriginal people were growing grain and storing it, thousands of years prior to European arrival.[12]

I'm a big fan of science. I have a bachelor of applied science myself. But we have to realise that science is a work in progress – that every discovery is at first guesswork and has to be proven over and over; confirmed, tweaked, rethought. Science is often very reductive, too, because it needs to control for outside factors; variables, they're called. We only have to look at glomalin to see that soil science has been playing catch-up for a while now. Prior to 1996, talk of such a substance would have been considered ridiculous. The realm of quacks and charlatans, according to the wisdom of the era. As some undergraduate students joke, the real leaps in science only happen as the graveyard fills, because every generation has their agendas, biases and blind spots. Sometimes we can only move forward when the leading scientific minds of one era are no longer with us.

The problem comes with separating the weird and mythical from the possible and provable. With soil, however, a lot of humans have had a lot of experience of it, first hand. For a very long time, we were all hunters and gatherers, then foragers or farmers. We were tied to the land, and our fortunes matched those of the soil on which we walked and worked.

Things are different now, and science has given us the ability to farm more of the Earth's surface than ever before. For over a hundred years, modern, intensive agriculture ignored the anecdotal in favour of the reductive. Where once we were all pretty much farmers, or very close to the land, and we rotated crops, studied yields, used livestock as land fertility aids and worshipped soil gods over lifetimes, not just financial years, now we just ring up the agronomist and order NPK. Where we once all felt, in our hands and backs and stomachs, the health of the soil that nurtured us, now we look at spreadsheets and averages and statistically relevant results. For a while, we lost the gratitude our forebears had for the soil they worked, the land they trod, the wisdom they garnered by growing stuff and watching the earth with more than passing interest, and more of a survival instinct.

There will always be a showman, a snake oil salesman, somebody who wants to flog you something so they can turn a dollar. But there are also visionaries who see the world in a new way, who can help us grow the food we need while also caring for the soil.

So, we've seen how grazing can trigger more plant life, above and below ground. Grazing means there are more exudates for our wonderful underground community, the microbes – and they, in turn, generate more grass. If soil benefits from grazing, and it grows food for humans to eat, that's probably not a bad use of land. But can it really reverse desertification?

Over 30 per cent of the world's landmass is grassland,[13] variably called savannah, meadows, rangelands and pasture. A third of that is in danger of turning into desert.[14] If we get it wrong, it can go really wrong.

However, there is a way to grow food that doesn't wipe out the soil. 'Regenerative agriculture' is a movement sweeping the globe. And while rotational grazing is often part of it, regenerative agriculture doesn't just involve moving ruminants from place to place; it also looks at the farm from a new angle. As we'll soon see, soil doesn't care if we want to eat animals or not. It doesn't care if we want to eat wheat or turnips or strawberries. What soil needs can be broken down into a few basic truths that are as useful to an American corn grower as they are to a New Zealand market gardener and a Welshman with an allotment.

CHAPTER 20

A Grain of Truth: Regenerative Agriculture

In March 2020, as the globe was headed for a COVID-19 lockdown, I spent a few days driving through regional New South Wales. In recently drought-affected country, I saw kangaroo grass standing over a metre tall along railway sidings and road edges. I witnessed the power of water to transform what had been dust bowls just two months prior into emerald-green fields.

But then, in the midst of all this crazy growth, all this sucking down of carbon into the soil by living plants, where the microbes were getting a long-awaited feed, I saw a graveyard. A graveyard of plants. Up towards Tamworth I passed thousands of hectares of land barren of life, grasslands killed by glyphosate, ready to sow crops – canola, wheat, sorghum, sunflowers and barley. The land was ready for no-till farming, the sort where a tractor direct-drills the grain seed into the ground with no need for the plough.

It took my breath away. So much land is sprayed to produce oil, grain, and sometimes cotton. So much land taken out of the act of photosynthesis for a time, even though this kind of growing drastically reduces erosion compared to conventional agriculture. And it does help reduce the need for fertiliser. And it does slow the loss of carbon, compared to tillage.

Soil is amoral. It wants and needs things that have nothing to do with our human propensity to form beliefs, such as a belief in organics or in glyphosate. Soil isn't interested in the politics of it, in the tribal wars of conventional versus no-spray. From soil's perspective, herbicide use and no-till is probably the lesser of two evils when compared to ploughing, as fungal hyphae remain intact, if unfed for a while. The soil's underside is preserved below ground, releasing less carbon than it would if it was exposed and the microbes left to die. And the mineral side of the soil – the sand, silt and clay – is far less likely to wash or blow away.

But seeing browned-off fields did make me wonder. If there's so much pressure on cattle farmers to prove they are sequestering carbon, storing carbon underground, as many require of Allan Savory and his disciples, what of this land? It, too, is vulnerable, generally losing carbon and structure.

Can we grow all kinds of food, grains and vegetables included, in a way that doesn't wreck our soil – and if so, what will it take?

When you delve down into the realities of growing food, it's inevitable that we cause damage to the natural world. What matters is less about the damage that is caused in order for us to live, and more about what is *sustainable*. Can we continue to do what we do forever? In other words, can the damage we do be repaired by nature? Pick a time frame – it doesn't have to be a year, and in soil terms, a decade may be more practical. Sustainability, in its truest, most basic form, is a cycle of damage and repair.

Compromises are constantly being made, either over time, or over a physical space, to get human food from the land. You'll do it yourself when you dig spuds, perhaps knowing that the all-important structure of your soil is ruined every time you dig them up. Every time you harvest annual crops from your backyard garden, there's the chance that you harm soil and remove nutrients.

But sustainable growing can be done. Chinese terraced fields have shown it to be possible. Can we do even better than that?

Can agriculture go beyond sustainable?

Can growing food actually lead to healthier soil than before we humans became involved? Higher in carbon, higher in microbial life? Better at retaining water, with a better structure, and ultimately producing food from healthier soil – food that is better for us to eat?

The answer – unequivocally – is yes! *Terra preta* and African Dark Earths have already proved what we can do, if we put our minds to it.

In farming terms, the new movement is called regenerative agriculture. It's not just sustainable, which implies no negative change over the long term. It's *regenerating* landscapes, replenishing water tables, supporting ecosystems, and fostering subterranean life in all its forms.

Well-managed grazing can increase carbon in soil, and increase biodiversity.[1] But it's far harder to do that with grains and vegetables unless we change the way we grow them, because both are usually extractive processes. Cattle can eat grass, store carbon, make milk and meat for us, and do it on land that isn't useful for growing crops. Restoring areas to pasture is also one of the most commonly recommended ways to restore degraded arable land from the damage caused by growing grains and other crops.[2]

Grazing may not feed soil as well as a forest might, depending on where you live – but it can feed more people than most forests. According to the American geologist, soil lover and author of *Growing a Revolution*, David Montgomery, if we wanted to just grow forests and live like we did 10,000 years ago, before agriculture, foraging and hunting, we'd need about 500 Earths.[3]

Having access to another 499 Earths is not likely to happen anytime soon, so we need to look at how we farm on the one Earth we have, and how we can do it better.

Regenerative agriculture looks at farms in a holistic way. Not quite in the 'farm as organism' kind of way that Steiner proposed, but not dissimilar. It recognises that everything is interconnected. The soil, the plant, the animal, the atmosphere, the water – they're all bound together. And throughout them all is this invisible ecosystem, the microorganisms that were here before us, and into whose world we have all been born. A hotbed of birth, death, alliances and enemies, all happening under our feet, in soil.

Regenerative agriculture looks at native biology, as well as the farm ecosystem. It considers lots of trophic levels, the feeding levels, from plant down and soil up. Growers who think regeneratively consider the consequences of wind, of soil carbon, of plant species and variety, as well as the consequences of livestock. Sometimes the best use of land, the best way to feed soil, involves the use of a domesticated animal.

Regenerative agriculture has five key components that align with all soil health recommendations, no matter where you live, or in what climate, on what kind of crushed-up rock, and what you're growing.

In short, the five principles of regenerative agriculture[4] look like this:

1. Keep the soil covered (no bare earth).
2. Minimise soil disturbance (don't dig).
3. Aim for diversity (in plants and animals).
4. Make sure you have living plants all year round.
5. Integrate livestock.

Some suggest a sixth principle[5] (which may actually be at the top of the list) – that every farm should be treated differently. Every patch of land isn't the same. No two farms, no two seasons, no two soil types are identical.

The first four principles are relevant to growers, be they producing peaches, tomatoes or sheep. The last one, livestock integration, isn't absolutely essential, but using animals in the system certainly doesn't hurt. There's no natural ecosystem on land that doesn't have animals

214

in it, so using them, wild or domesticated, or their by-products in your home plot, is a good way to go.

There is huge potential in regenerative agriculture. Remember, while plants are the origin of our energy source, they are only part of an ecosystem. They can and will draw down carbon, and the best place for that carbon to end up is in soil, because there's more potential for soil carbon storage than there is in plants themselves, which are more easily burned, eaten or desiccated.

Humans now know how to measure carbon. We know that more carbon in soil begets more life, more moisture, more vigorous plants. More life begets more resilience. More resilience and more life begets better nutrition.

We know that animals, when used as part of a soil-building system, cut the time for organic matter to be broken down, and can build carbon in the soil.

But livestock have another impact – and to understand that, we first have to consider the climate.

CHAPTER 21

What's the Beef
with Methane?

It's early September and outside my office window lies a paddock, fringed with silver wattle, bursting forth in millions of golden blossoms. Behind the silver wattle, a pioneer species, lies a forest of stringybark and peppermint eucalypts, which once dominated this landscape. It's early spring and a warm wind is gusting up the gully. The grass has turned emerald green in the paddock, overshadowed by wisps of brown. The thin stalks left over from last year billow like a bald man's comb-over.

If we stopped farming, this land would probably revert to eucalypt forest. But not the same forest, because that was managed land; the original inhabitants, the Melukerdee people, burned in a patchwork pattern to create open woodland. Aboriginal people weren't arbitrary spectators in the use of this land; they altered it to feed themselves.

We alter it, too, and grapple with the sometimes-competing needs: to look after soil and still produce nutritious foods, as all farms should. Looking after soil, we think, can include grazing animals – which for many is controversial these days.

If there's one area in which the battle for the food dollar has met the battle for the climate head-on, it's meat. In particular, meat from grazing animals, where those who believe we should all abstain from eating meat have found the ultimate enemy – ruminants, livestock whose digestive systems release methane, a potent greenhouse gas.

The single best thing you can do for the environment is stop eating meat – according to Joseph Poore, the co-author of a scientific paper, 'Reducing food's environmental impacts through producers and consumers', published in *Science* magazine in 2018.[1] That's a quote that made headlines around the world, including *The Guardian* newspaper, and has been used to energise plant-based food proponents.

On the release of his scientific paper, Poore came up with a misleading – and unfortunately often repeated – statement: 'A vegan diet is probably the single biggest way to reduce your impact on planet Earth, not just greenhouse gases, but global acidification, eutrophication, land use and water use,' he said. 'It is far bigger than cutting down on your flights or buying an electric car.'

Problem was, his review did not look at anything other than food. Not coal. Not fugitive emissions from fracking or natural gas extraction. Not petrol, or the embedded energy in steel or concrete. It didn't look at what's possible in soil. It also used a fairly arbitrary carbon life-cycle accounting system that allows for emissions from livestock, but not carbon cycling in pasture, and the earth that grows it. Emissions, but not capture.

Hearing words to the effect that 'the best thing you can do for the climate is to avoid eating meat' set the anti-livestock movement in motion. Finally, their cause wasn't just about the exploitation of animals. Or about selflessly giving up what has long been for many a dietary staple. It was about saving the world.

Animals are part of just about every functioning ecosystem, and humans have been altering ecosystems for a long time in order to feed themselves. The longest-running civilisations all knew how to regenerate the landscapes that nourished them. Much of what grows in our landscape is grass.

Grass is made mostly of cellulose, a fibrous form of carbo-hydrate, and cellulose has the largest mass of any carbohydrate on the planet. It's over half the carbon found in the vegetable kingdom. Cellulose is, sadly, an indigestible starch to us – but not for grazing animals. Horses can digest cellulose, as can kangaroos, but both roos and horses have single stomachs and are inefficient digesters of cellu-lose, which means they need to eat at least 30 per cent more grass to extract the same amount of energy from it as ruminants.

Ruminants are a cluster of animals that share a similar style of digestive system, one that is more complicated than ours, and more complicated than a horse's. Ruminants have four stomachs, the largest of which is the rumen, which gives the grouping their name. Ruminants include cows, sheep, goats, deer, elk, moose, giraffe and camels.

The reason ruminants are more efficient at digesting cellulose boils down to microbes. A lot of things in this book, you'll find, boil down to microbes.

Ruminant gut microbes can ferment cellulose and turn it into energy that the animal can use. But even more importantly, as those microbes ferment and consume the cellulose, they turn into dead microbes that provide the animal with protein. (Between 70 and 90 per cent of the protein that a cow gets in its diet can come from dead microbes.[2])

The rumen in a cow has a quadrillion microbes[3] – a thousand million million microbes, or a thousand trillion, if that helps. These microbes are mostly bacteria; some are archaea, a single-celled, ancient branch of life. Some archaea species can readily digest cellu-lose, and in the process they release methane. Archaea are the only microbes known to be *methanogens*; that is, able to produce methane.

When it comes to including livestock in a managed ecosystem, methane is a glaring problem. Methane is not just any greenhouse gas, it's the second most important one after carbon dioxide.

No respectable climate scientist disputes that climate change is real. (And, let's face it, consensus is mightily rare in science, because what young scientist isn't trying to undermine their elders with a better theory?) But let's look at some of the greenhouse gases, and soil's role in them.

Soil, as we have seen, is made up of rock particles, living things and formerly living things, plus water and air. Now the air bit is really astonishing. While we like to think of the earth under our feet as quite solid, it's actually very much active and porous. Air moves freely in and out of it. Good soil can be up to 25 per cent air. That's right: a quarter of soil can be air. Who knew?

What's more, if your soil is well drained, the air in the top 20 centi-metres of soil is completely renewed every hour. That's right, soil breathes! It has been estimated that each year, 30 per cent of the world's carbon dioxide, 70 per cent of the methane, and 90 per cent of the nitrous oxide released to the atmosphere passes through soil.[4]

This shouldn't come as a surprise. All those microbes, they are living things. Most are reliant on oxygen, and release carbon dioxide, in the same way we humans need oxygen and release carbon dioxide. But there are other soil microbes. Some, such as some euryarchaeota species, can be methanogenic. These microbes live in oxygen-depleted soils – such as flooded land, like rice paddies – and actually take in carbon and hydrogen, and emit methane.

Methane gets a bad rap in the climate change stakes. Not surpris-ing, really, given it has a more powerful warming effect than carbon dioxide. Until recently, methane was estimated to be about 28 times more warming over 100 years than carbon dioxide, based on theo-retical models (ones Joseph Poore had been using). But that 28 times more warming than carbon dioxide has been revised down to eight times more warming by looking at actual results more closely.[5] Still, eight times worse is eight times worse; it's just less bad.

For a long time, the things that emitted methane – such as the euryarchaeota, and the cows – were in balance with other things. As cows and other ruminants emitted methane, mostly through burping out gas from their rumen, and wetlands and bogs and rice paddies emitted methane from soil, there was a natural harmony. That's because there are things called *methanotrophs* living in – you guessed it – soil; a group of microbes that actually digest methane and use it for energy. These microbes, which have so far eluded most efforts to cultivate them in the lab, can thrive in the earth. In healthy soil.

A thing that can help moderate the climate effects of methane lives in soil!

In one experiment from 2019, published by the US National Academy of Sciences,[6] researchers point out that while most methane is broken down by sunlight in the atmosphere, the only living process by which this occurs – the only one we have any control over – is through soil.

The scientists went on to isolate a methanotrophic microbe called *Methylocapsa gorgona*, which has a super-strong affinity for methane, meaning it can and does draw down methane when the gas is in the atmosphere.

Ruminants evolved in the presence of microbes, and have been around for a very long time (they started outcompeting other herbivores 20 million years ago)[7], so the soil biome and the ruminants have co-evolved. It's hardly surprising, then, that the bacteria in the soil underneath where the cow is burping are busy converting that methane into more harmless components.

While it's a hard area to research, and it's very early days, it seems the methane-eating bacterial colony expands when there is more methane directly above the soil, and that a ruminant-grazed pasture can support more methane-eating microbes.

Studies on methanotrophs have shown, consistently, that they exist in higher numbers in healthy soil, and that deeper soil has more methane eaters in it.[8] But in places where soil is compromised, or where artificial fertiliser is used, the methanotrophs drop – not only in number, but also in diversity. Studies also show that the converse is true: improved soil health leads to an increase in methanotrophs, in both number and diversity.[9]

In other words, if you want to get rid of methane, don't use artificial nitrogen – use natural fertilisers and let soil health take priority.

Methane in the atmosphere is generally consumed within a decade or so,[10] mostly by a radical reaction in the troposphere, the lower atmosphere. A short methane lifespan is a good thing for the climate.

What that means, in a practical sense, is that if you have a set number of cows, or rice paddies, and don't increase those numbers over ten years, then you don't add to global warming. The methane your cattle or goats or sheep or rice paddies add to the atmosphere today is gone in a decade.

So, in theory, if you don't add to the world's ruminant herd, or flood more land to grow crops, then the methane in the atmosphere should stabilise, thanks to the troposphere, and methanotrophs.

But the thing is, the amount isn't stabilising. Atmospheric methane has been increasing by 10 million tonnes a year, or about 1 per cent, on average, for the last two hundred years.[11] Much now comes from fracking for natural gas. Some comes from traditional gas drilling and oil wells. Some is coming from thawing tundra, melting ice sheets, coal mines, burning off, rubbish tips, sewage plants, slurry pits and drying peat bogs. Some comes from growing rice, of course, as does a proportion from ruminants.

It appears, however, that recent increases have nothing to do with the ruminant herd, or the rice paddies. Despite no real increase in the global ruminant herd between about 1990 and 2010, methane kept rising at the same rate it has since the Industrial Revolution.[12]

The culprit, as ever, is the burning of fossil fuels. At the same time, we're destroying the only way we know to actively get rid of methane on Earth – soil microbes.

Soil microbes currently degrade up to 20 per cent of the methane that is produced, but they can only do that where soil is healthy.[13] And they could break down much more methane if land is fertilised using compost. Methane-eating microbes work far, far better if you let grazing animals poo on the grass that they graze, not lock the animals away in intensive farms and bring in food for them from elsewhere.

You won't be surprised to hear that we don't really understand methanotrophs in soils. We have no idea how many thousands of species exist. We have evidence they are there, but we can't identify them. We do know that heavy grazing decreases the ability of methanotrophs to work, as does soil compaction. But evidence from 2020 shows light grazing can increase the methane-digesting ability of methanotrophs by over four times, compared to ungrazed grasses.[14]

It's also now known that increasing levels of soil carbon, which is fast becoming a focus of this book, usually increases the number of methane-consuming microbes the land contains. In contrast, ploughing land diminishes the ability of that piece of earth to absorb methane.

In other words, grazing land is good for methane reduction. Cropping land isn't, unless it's no-till: research has estimated that no-till agriculture can help break down methane up to 11 times faster than ploughed land.[15]

Soils under rotational grazing systems work even better at removing methane than no-till cropping. Rotational grazing also emits less methane and nitrous oxide.[16]

Can soil methanotrophs under a single grazed paddock actually neutralise the methane emitted from the animals chewing their cud after eating the grass that grew on the same land? It's unlikely, but they don't need to – because it turns out that forests are usually better at oxidising methane than grasslands. So as long as we keep wooded areas near our grazing land, they will do a fair amount of the work, too.

I'm not saying you have to eat meat; I covered the ethics of that in my book *On Eating Meat*,[17] and you can make your own mind up about how to best frame your diet. What I'm saying is that we should consider what soil needs if we want to feed ourselves into the future.

Regarding emissions, however, there's another thing to note. Methane emitted by a ruminant isn't new carbon added to the system. Carbon, as we know, is stored in oceans, in soil, in the atmosphere, as well as in living things like us, and microbes. Some carbon was stored 300 million years ago, during the Carboniferous period, when masses of plants and algae grew and were compressed underground. The sugars those plants created out of the air, which is long locked-away carbon, are now our fossil fuels.

The methane a cow breathes out isn't from that source. It's made up of what the cow ate. If it ate grass, some carbon is fermented into energy and protein for the cow, and some is resynthesised and released as methane. This carbon isn't the same as fossil fuel emissions from a coal mine, a gas well, or even carbon that's been trapped in permafrost. It's a short-lived gas, removed from the atmosphere and put back into the atmosphere, as it cycles from the air, to soil, to plant, to cow, and back to soil, then the air.

Carbon released through fossil fuels, however, is a one-way street. Carbon laid down millions of years ago is burned as coal, oil or gas – and released as new carbon into our modern atmosphere.

This isn't to shy away from the very real problem that absolute levels of global methane have gone up by about two and a half times in the last 300 years. Methane is at the highest level it's been in at least 800,000 years.[18] As a higher-impact, short-acting greenhouse gas, many want to cut its output to give us time to deal with the actual culprit – new carbon being put into circulation by the burning (or fugitive emissions) of fossil fuels.

If we killed every ruminant on Earth today, we'd get an almost immediate global cooling.[19] However, this effect would give us only about ten years grace before we were back to where we started when we killed them – without any of the benefits grazing animals can provide.

One of the main drivers of the anti-ruminant debate has nothing to do with soil, or the methane emissions of the current herd. Most of the research is taking a long-term view, watching how, globally, our meat consumption is rising.

Globally, annual meat production is projected to rise from 218 million tonnes in 1997–1999, to 376 million tonnes by 2030.[20] A whole bunch of that increase will be pigs and chickens; perhaps sadly for soil, these animals eat grain grown on land that is notoriously bad at absorbing methane (while often emitting carbon). But some increase will be in sheep, cows and goats. Despite the global cattle herd decreasing for a while recently, the UN's Food and Agriculture Organization estimates that between 2000 and 2050, the global cattle population will increase from 1.5 billion to 2.6 billion, and the global goat and sheep population will rise from 1.7 billion to 2.7 billion.[21] This will, obviously, put more pressure on the methanotrophs and the climate.

Unless we can be sure that grazing can guarantee enough methanotrophs in the soil directly under where these animals live – and, also, in arable land – we probably should look at not adding substantially to the global ruminant herd, for the soil's sake, and the atmosphere's.

And that could mean that some of us in the world need to eat less meat.

Can we restore soils enough to cope with all the methane being emitted – not just from the mouths of cows, but also all that methane from swamps and rice paddies, and coal mines and fracking?

At the moment, it is our only hope.

Soil isn't some innocent bystander in the climate change debate. It contains all the elements that make up the most important greenhouses gases: the carbon and hydrogen from methane. The carbon and oxygen from carbon dioxide. And the nitrogen and oxygen from another highly warming greenhouse gas, nitrous oxide. Soil can act as either a sink or a source of greenhouse gases.

Perhaps surprisingly, the top one metre of the Earth's crust has way more carbon in it than the air. Soil, both the living and decomposing lives within it, contains 4.5 times more carbon than the rest of that biosphere put together – including all the plants and animals on Earth.[23] All that carbon was, or still is, the result of living things.

In other words, while we often view trees as a carbon sink – a way to store carbon that's not in the atmosphere – it's actually soil that is doing the heavy lifting.

In fact, according to the American author and geomorphologist (landscape geologist) David Montgomery, Professor of Earth and Space Sciences at the University of Washington, about a quarter to a third of all the new carbon added to the atmosphere since the Industrial Revolution is from soil.[23] All that glorious organic matter, that wonderful liquid carbon pathway that Australian soil scientist Christine Jones talks about as plants draw down sugars into the soil[24] – that can also be released.

As we've eroded, overgrazed, cultivated and impoverished soil, we've allowed lots and lots of carbon to be emitted into the air. If growing food hadn't been so bad for soil, the increase in levels of atmospheric carbon since the Industrial Revolution would be about 35 per cent lower than we have today.[25]

Farmland is still releasing soil carbon as I write. But some regrowth forests, and well-managed grazing lands, are helping soil to actually be a net sink of carbon.

In a good news story, for the last two decades we've actually increased the amount of carbon stored in plants and soils, by better managing some of our land.

As we've seen, carbon in soil is stored in plant roots, in microbes, and in soil organic matter. There's short-term storage in manure, urine, living microbes and finer roots. There's medium-term storage in humus and lignin, in sturdy roots of trees. And there's long-term carbon storage in glomalin, biochar, organic compounds and some other fractions of humus.

Of course, there are farmers – mostly, but not exclusively, regenerative and organic growers – who are actively increasing soil carbon. And the evidence is that we can sequester, or store, more carbon in soil than we thought, and more than we have been. Way more.

In fact, the ability of soil to store carbon is one of the climate's great hopes.

According to the leading figure in the field, Rattan Lal, a soil carbon expert who was awarded the 2020 World Food Prize, the total carbon in terrestrial ecosystems is approximately 3170 gigatons.[26] (A gigaton is big. It doesn't matter if you can't quite work out how big, it's the relative numbers here that matter.) Of this, nearly 80 per cent (2500 gigatons) of carbon is found in soil. By comparison, the amount of carbon in living plants and animals is about 560 gigatons – only about one-sixth of the carbon that soil stores.

This also means there's three times as much carbon in soil than in the air (800 gigatons) – and that's after we've released heaps of carbon into the atmosphere, during the last hundred years in particular.

The ocean is the largest carbon pool, at about 38,400 gigatons, but even then, at current rates, the ocean is only storing carbon at a rate of about 2 gigatons a year, compared to the soil's 3 gigatons a year. More carbon in the ocean tends to lead to water acidification, a bad thing – whereas more carbon in soil is pretty much always a good thing.

So, despite the depleted carbon levels in our modern-day soil, after a hundred years of fossil fuel–driven tractors, centuries of ploughing, and decades of artificial fertiliser use, soil is still a big carbon sink.

What's more, it can be an even bigger sink. How big? Well, according to former French agriculture minister Stéphane Le Foll, it can be such a big carbon sink that it can heal the world.

In 2015, when the world's leaders met to discuss solutions to climate change at the Paris Climate Summit (home of the Paris Accord, where global leaders committed to reducing emissions to protect against runaway global warming), Le Foll touted a simple idea for helping pave the way (or should I say, *garden* our way) to planetary health.

In French, it's called *4 pour Mille* ('4 per 1000') – and it relates to a percentage increase in soil carbon. If the top 30–40 centimetres of all the world's agricultural soils, and soils under human management, were to increase in carbon by a tiny amount, by just 0.4 per cent a year (which is 4 parts in 1000), virtually the entire global increase in carbon emissions for each year could be offset.

If we did this for a few years, while we transitioned out of fossil fuel use – the *real* culprit in the long term, and something that simply has to be addressed – we'd not only improve soil, we'd also be helping the climate.

How does that work? Well, we know human activity is increasing the amount of carbon dioxide in the environment. Only about 30 per cent of the carbon that is released each year is currently taken up by plants as they photosynthesise. And of course, plants and soil work in tandem. So, all the carbon that we met in previous chapters – all the sugars synthesised by plants from thin air, all the glomalin trapped by fungi in soil, all the living beings, like microbes and so on, that are made predominantly from carbon, and all the dead, formerly living matter, including humus – these are all made up of the *same* carbon that is causing our climate woes. It's the same stuff, in a different form. And while we may not be able to put back into soil *all* the carbon that it has lost through our ignorant treatment of the earth, we *can* put a whole heap of carbon back in.

Just a tiny fraction – a minuscule-sounding 0.4 per cent increase in soil carbon per year – can offset so many other emissions.

Even better, as we have seen, an increase in soil carbon does other equally vital things. It allows the soil to store water better in dry times, and to drain better in heavy rains. It holds the soil together better. More carbon in soil means more methanotrophs, digesting methane. And, of course, more carbon, more of all the other subter-ranean life, which means more micronutrients that are available to us if we eat those plants.

It's a win, win, win, win.

The *4 pour Mille* agreement, initiated in France and signed by the majority of nations on Earth, acknowledged the power of agriculture. Where the focus on carbon sequestration (storage) has often been solely on trees, or large-scale carbon capture by humans at power plants (which is yet to be proven viable), '4 per 1000' talked about the massive sink we already control. We just have to use soil in the way it's intended.

Storing carbon in soil was not on the public agenda in 2015 when Le Foll first mooted his scheme. Things have changed a lot since then. In 2020, fossil fuel giant BP announced it is working in the soil carbon space (probably, the uncharitable might argue, to offset its emissions). Bayer (formerly Monsanto, most famous for their glyphosate herbicide Roundup) is paying farmers to store carbon in soil by promoting a system of spray and no-till, which isn't completely ideal, but is at least a start. The Australian government, even under a series of remarkably denialist leaders, have started paying farmers to store carbon in soil. Meanwhile, Ireland's carbon accounting makes specific mention of the role of grasslands and their carbon-storing capacity in their response to the climate crisis.

4 pour Mille is an ambitious target. You'd have to get every farmer and every land manager on the globe actively working to store carbon. Mono-cropping grain growers and vegetable producers would struggle to store carbon at that level. Many graziers might be hard-pushed to produce such results. Not everybody has to achieve the same level, but the average, across all human managed landscapes, is a big ask.

However, it *is* possible, and it is happening in some places. Regenerative agriculture is just that, and its popularity is skyrocketing. The big problem is mostly political: half a decade after the Paris Climate Summit, only 37 of about 200 countries have committed to storing an equivalent amount of carbon in their soils.

Le Foll's vision is still alive. The *4 pour Mille* movement continues, and points the way to a more secure food system. While it has powerful detractors, it also has powerful allies. Much work can be done in this space, with huge benefits to the environment.

There are people all around the world working on ways to store carbon underground. And as we'll see, one of those leading the way is combining scientific knowledge with farming know-how and harnessing the ability of the best carbon pumps we know of: plants.

NO LAUGHING MATTER

The third most important greenhouse gas is nitrous oxide, laughing gas, representing 20 per cent of all human-induced global warming.[27] While carbon dioxide has risen 50 per cent since 1750 (just prior to the Industrial Revolution), and methane has increased 250 per cent in the same time, nitrous oxide levels have risen 20 per cent – from below 270 parts per billion (ppb), to more than 320 ppb.[28]

That doesn't sound like it should worry us, except that nitrous oxide is about 300 times more harmful to the climate than an equivalent amount of carbon dioxide.[29] And unlike methane, it isn't short lived. Nitrous oxide can cause acid rain, and it also destroys ozone, the atmospheric layer that helps protect the planet from harmful ultraviolet rays.

As we saw earlier, the largest emissions of nitrous oxide come from farmers or gardeners spreading artificial fertiliser. In fact, it's estimated that 63 per cent of the nitrous oxide that the United States produces comes from fertiliser application.[30]

What can soil do to help alleviate nitrous oxide? Well, not a lot is known – but globally, about 70 per cent of nitrous oxide in the atmosphere comes from soil.[31] Some of this nitrous oxide is a natural by-product of normal biological processes.

But we know that if you want less nitrous oxide in the air, the same rules apply as for methane and carbon dioxide. What you need under the ground is living plant roots, less soil compaction, more air, and a more vibrant ecosystem.

So don't spray with herbicides, don't use artificial fertiliser, and don't plough the earth. If you must dig the ground, adding biochar to the soil reduces both methane and nitrous oxide emissions, while also storing carbon for the long term.

Every little bit helps. The *4 pour Mille* agreement relies on lots of small and large changes, on all human-managed landscapes, from the biggest corn farm to the smallest rose garden, recognising that tiny changes in how we tend our land can have a huge impact on the earth that nourishes us.

CHAPTER 22

Money in the Bank

If you were a grower, and I told you there was one thing, above all else, that would make your land better, and that you could grow more food on the same land, using the same rainfall, on the same earth that you currently steward, using less inputs – meaning it would cost you less – would you do it? You'd probably think I was a con artist. Such a thing, surely, isn't possible?

If I told you someone might *pay* you to do it, you'd probably laugh in my face.

But, that's the truth of it. There is one thing, just one thing that matters more than any other, that you can do for your soil.

This is even despite the complexity of soil, the interplay of the physical, biological and chemical. Despite the seemingly countless players in the soil food web – their thrust and parry, and the innumerable interactions creating thousands of individual chemicals for plants to use. Despite, even, the flurry of new science starting to decode microbes and their specific functions in finding food for plants, aiding communication and protecting against disease.

Despite all of that, there's just one thing to focus on, and it's an element we've touched on before: carbon.

Farmers have lots of clichés. We've given the world 'make hay while the sun shines' and 'kicked the bucket' and 'common as dirt' and 'like a pig in mud', as well as myriad plough sayings. But, within the farming community, there are a few more: 'hay is money in the bank' and 'mud is money' and 'reap what you sow'. They reinforce the idea that stored summer sun, in hay, is like cash, that rain brings a good harvest, and that labour equals success.

When we, as humans, consider soil, we think, what can it do for us? Which isn't necessarily a bad thing. How much grain can it grow for us? How many goats will it support so we can make cheese? Where can I plant trees to build my house and protect my garden?

But as we've seen, soil is in limited supply. There's another saying in farmer circles: 'Funny thing about topsoil, they're not making it anymore' – which, it turns out, isn't true.

We've seen how quickly topsoil can be lost, through erosion, desertification and simple ploughing, and we know that only about 7 per cent of the world's surface is available for growing food[1] – and we've built on a whole heap of that, too. Sadly, despite its miracle qualities, there's not enough compost in the world to nourish all our agricultural land and maintain high yields.

We also know there's no Planet B, and that we're turning a lot of our current growing country into a moonscape.

But, the wonderful thing is, there's a whole world of clever people working out how to build topsoil. Not at time frames nature might take; 1000 years to make a centimetre or two. And not taking even a human lifetime to build soil. It can actually be done in far less time than that, if we put our minds to it.

We've seen how humans in the Amazon and Africa built soil with their Dark Earths. How people built topsoil in Holland with *plaggen* soils. It's also known that Aboriginal people in Australia created a more verdant soil with low-temperature mosaic or patch-work burning. Humans have always managed land – and in some cases, have improved soil and grown crops for aeons.

But, overall, we've never treated soil more badly than we have today – yet at the same time understood it more. We need to make soil faster, because we've ruined it much more quickly, and on a far broader scale, than at any other time in history.

So, the central question returns. How do we make soil?

First, a quick recap.

Remember what makes soil, soil? It's sand, silt and clay that has a biological component, a living part. It's also only soil when it's in association with plant roots, or it has recently been in association with plant roots. And it needs air and organic matter that isn't living. And moisture to stay alive.

As we've seen, sand, silt and clay aren't running out any time soon. They might be present in the wrong proportions, or the wrong place, but generally we can deal with that on a global scale if we stop eroding soil. There's plenty more rocks where those ones came from.

The biological bit, however, is woefully lacking in much of our growing land, and our soils are becoming more lifeless.

So how do we fix soil? The simple answer is carbon. The same carbon that was emitted from the soil, thanks to the use of artificial fertilisers and the plough, is the very thing that is missing. Without carbon, the soil microbes don't have much to eat. Without carbon, the soil's structure is poor, unable to hold air, or water. Without carbon, we can't support fungal hyphae, worms, nematodes and collembolans.

Put more carbon into the soil and the biology will follow. And the carbon we want to put back into the soil is the same as we saw right at the start of the book – it's the same carbon that is drawn down by plants. It's the same carbon that's rising in concentrations in the air right now, the carbon dioxide that is contributing to global warming.

Carbon out of the air, and back in the ground, is what we need to heal the world.

While some scientists have been genetically modifying corn, perfecting herbicides and breeding higher-yielding wheat, other scientists and growers have been working out modern ways to make topsoil, using biochar, compost, nitrogen-fixing plants, cover crops, and even nano-clays that can help sandy desert soils store moisture.

All of these have a place, and a time.

But the original method is still the best. A plant is like a permanent, low-labour, fossil fuel–free way to suck carbon from the air and pump it underground. That's what plants do. Plants can only do it efficiently, however, when we focus on soil – not just the plants alone, or the animals that eat those plants. Otherwise we get sidetracked and forget the main aim.

As we've seen, plants love diversity. They don't necessarily compete for resources; in fact, they often help each other. Because of quorum sensing – the ability of microbes to kick into gear once they sense the presence of enough other microbes – diversity in plantings can help. If you sow multiple species in an area, nature can thrive.

Most plants pump a lot of carbon down into the ground in the first part of their life. In fact, if we work with the growth cycle of multiple species of plants, we can manipulate the carbon cycle in soil's favour. We can use sunlight – that copious, underutilised energy of the sun – to pump carbon straight into the soil. What sunlight gifts the globe *each hour*, in terms of energy, is the same amount as we burn *each year* in fossil fuels.[2] A year's worth of energy falls on Earth every hour. For free.

It's an infinitely renewable resource that we can use better, for no cost, and only benefit.

All of this leads us back to Niels Olsen. Niels is a beef farmer, machinery tinkerer and soil nerd. We met him back in Chapter 2, when I first discovered soil that looked like chocolate cake. Niels is actually getting *paid* to store carbon in soil. In a global economy that is tentatively trading in carbon credits, Niels was the first person, anywhere, to show that you can store carbon in soil in a rapid, measurable, verifiable way – and be paid by a government to do it. He was also the second person to be paid by a government to do it because he was so far ahead of the field and no-one else had done it in the meantime, though he's very keen for others to follow in his footsteps.

First, some numbers. Testing in 2018 showed that soil organic matter in Olsen's soil rose from around 3 per cent to 10.7 per cent in just five years. He now has enough plant-available phosphorus to last 600 years. The farm's worm and microbe populations are now six times bigger than they were just half a decade ago. Tests of his soil show improvements that look too good to be true: 34 per cent more plant-available phosphorus, 51 per cent more sulphur, and 122 per cent more plant-available nitrogen. And the soil is warmer, too, by about 1.5–2°C,[3] increasing the length of the growing season over the cold months in his part of Victoria, in southern Australia.

How did Olsen achieve such results?

In the late 1990s, Niels Olsen had a cracker couple of years farming and wanted to come up with a way to pay less tax. (I can hear a lot of farmers almost choke on the thought: 'Imagine earning so much that you have to avoid paying taxes!'). So he decided to spend up on ways to make the farm more productive. Making the farm more productive for the long term, future-proofing it, is money in the bank. His agronomist, the industry scientist who advised him on all things agricultural, said he should put fertiliser on the soil. Niels agreed. If one bag per hectare was going to boost soil fertility and increase the grass cover, then two bags per hectare would be better. So he added a blend of urea, triple super, and muriate of potash, at rates two to three times higher than he had in previous years. And boy, did he grow grass. Tall, lush, bright green grass. He thought he'd invested well.

But the grass all died off quickly in the first heat of summer. Olsen called in the agronomist, who repeated the advice. So Olsen obliged and did the same mix again. And his results got worse.

Olsen discovered what generations of farmers have often seen: that quick, simple fixes can send your soil backwards. The Moron Principle, as Colin Seis, the pasture cropper we met earlier, calls it. Olsen just did it in a much shorter time frame than most.

Unimpressed with these results, Olsen decided he needed to know more about what went wrong. Move over farmer, enter Niels the researcher. He studied assiduously. He focused more broadly on the chemistry of soil, and to some extent its physical nature. He started applying different amendments and minerals to his paddocks to improve the land. He discovered the wealth of underground biology we've touched on in this book, and was fascinated by the notion of quorum sensing in soil. He toyed with methods of no-till agriculture, particularly pasture cropping. Being a practical bloke, he was also challenged by the way no-till machines planted those seeds without digging the earth completely.

To date, his life's work is embedded in an ingenious machine – a sort of combined slasher, seeder and hoe – that Olsen calls the Soilkee. The Soilkee attaches to the back of a tractor, and one pass over a field of pasture will sow a variety of seeds. Olsen now puts about 12 different species of annual plants in the mix, everything from tillage radish (a form of daikon), corn, vetch and peas, to plantains and barley. Essentially, it's like sowing a paddock with 12 different crops at once. But these crops aren't for humans to eat. They're all there to feed the soil.

This mix of plants creates a cascade of new life. The fungal networks spread and intertwine, the bacterial count skyrockets, and the plants suck up the sun's energy and funnel carbon into the ground. Some carbon, of course, is for their living roots. Some is for the microbes. This flurry of activity is at its peak in young plants – and you actually get better results if the crop is trimmed occasionally. Enter cattle.

Cattle, or other grazing animals, can trim the tops off the plants without burning fossil fuels. Gardeners might recognise a similar process in growing and cutting a green manure crop to feed their soil.

The cattle eat down the plants after about six weeks, and again at least once, sometimes twice, before the patch is re-sown six months later. It's a smorgasbord of food not only pumped into the soil by living plants, but also available from nitrogen-fixing microbes as well.

Masses of carbon – amounts previously thought impossible – are stored in soil this way. The audited results from nine independently verified plots show that the soil organic carbon increased on average 24 per cent in a single year.[4] That's a total annual increase of about

0.9 per cent – way beyond the *4 pour Mille*, the 0.4 per cent increase in soil carbon that was brought to the table in Paris. It's also the bit that makes all the other nutrients – the phosphorus, the nitrogen – more available to the growing plants.

This carbon also helps water retention. Put it this way: if you increase soil carbon from 1 per cent to 4 per cent, you can double the soil's water-holding capacity. You can also double the capacity of the soil to hold air.

As Australian soil ecologist Christine Jones puts it: 'one part of soil humus could, on average, retain a minimum of four parts of soil water'. What does that mean on the ground? Well, every 1 per cent increase in soil organic carbon, Jones says, 'equates to 168,000 litres of [extra] water that could be stored per hectare'.[5]

Increase soil's carbon level by 1 per cent, and you can store an extra 16 litres of water per square metre.

In. The. Soil.

Not in a dam. Not in a tank. Water retained on your land. More moisture ready for the next stage of plant growth, without needing a single millimetre more rainfall. It's what all growers, including home gardeners, strive for, so you can miss a day's watering and not come back to find desiccated vegetables. The exact same principle can be done with green manure and a scythe (or brushcutter) in a home garden.

The beauty of what Niels is doing is that it involves plants, not artificial inputs. It utilises sunlight to its best advantage, to nourish soil. It's an elegant, self-sustaining system that is also really simple. Run the sowing machine over the field and, hey presto, there's more phosphorus, more nitrogen, and even more water available in the soil. And all of it is driven by carbon.

All over the globe, growers have learned key lessons in soil health. The best growers now know that soil life is pivotal, and that soil ecology is not only driven by carbon, but also drives carbon.

Regenerative farmers, concerned home gardeners and others like them are taking the approach that we must look after the thing that looks after us.

Carbon, it turns out, is money in the bank – but only if you store it in soil. You can't increase the amount of land you have, without great cost. You can't change the component ratios of sand, silt and clay on your land without a great deal of effort (unless you can enlist a passing glacier). It's hard to add living microbes to your soil day after day, month after month, bought from people who now specialise in breeding them and pouring them on your patch. It's also expensive and labour intensive to bring in biochar and compost and mulch – not to mention costly in terms of fossil fuel use. These methods all work, too, and are useful in the tool kit. But they aren't the most efficient way.

Thankfully, there's this other way. It's called feeding soil. That's as true for the home garden as it is for a commercial goat dairy. With a little push from us, we can store carbon in soil far more quickly and efficiently than just waiting for nature.

While soil requires nature to make it – countless organisms and interactions, all acting in concert – we can act as the conductor of the orchestra. It just takes a change of focus, from solely on plants or animals, to putting soil first, and getting more living action in our soils.

CHAPTER 23

They Germinated a Seed on the Moon

A lot of things fill my head in the early hours while milking. As Bessie, or Elsie, or Myrtle, or a combination of our dairy cows quietly gift me milk each morning, I find myself pondering many mysteries. And one thing that has puzzled me for years set off a lot of the research for this book.

Outside our little micro-dairy, there's an anomaly. As I milk, I spy green shoots popping up in the patch straight behind the cow. Each girl stands on a concrete pad while being milked, her head held firmly in wooden struts so she, and I, don't get hurt from a sudden movement. After milking, the girls back out slowly, then meander off to find new grass. In order to make their backing out easy, I've filled up the space behind the concrete floor using what's known as 'fine crushed rock', which I buy from a local quarry. Essentially it's just what the name suggests, grey stone, a form of granite, some of it crushed to 1 centi-metre (half an inch) in size, most of it smaller. Some even closer to dust. It's not soil, though, because it's had no living plants in or near it at any time in recent geological history. It has no organic matter in it, either. It's rock and dirt.

The green shoots that appear in this rock are from a handful of barley that I use to sweeten the girls' rations each morning. Any spillage, a grain or two, plus the floor cleaning water, is swept into the ramp.

Any whole barley seeds are quick to shoot. They sprout forth in under a week, and can live for a fortnight, if the cows' hooves don't unearth the roots. Then they keel over and die. Inevitably die.

This cycle of life and death has me fascinated. How does a seed, in an area devoid of soil, kickstart its potential and stay viable for two weeks? My barley clings to life with no organic matter to speak of, and just rock dust as a home.

It turns out a seed is more than just a seed. A seed is a mini universe.

In January 2019, thanks to the Chinese Lunar Exploration Program, humans germinated a seed on the Moon.

The cotton seed was among six organisms taken aboard the Chang'e-4 robotic lander in a specially constructed 2.5 kilogram (5½ pound) biosphere, designed to replicate Earth's air pressure and atmosphere. The aim was to learn more about growing things in space.

The chosen organisms were a strain of yeast, seeds from rape (canola), cotton, potato and *Arabidopsis thaliana* (thale cress, a brassica, related to cabbage), along with fruit fly eggs. (Coming from an area of the globe that is free of fruit fly, why you'd take those eggs to a fruit fly–free region is a mystery to me.)

The only thing that grew, however, were the cotton seeds. They sprouted forth, and put out leaves before keeling over in nine days. They probably froze to death, as the lunar night can get to –190°C (–310°F), and the biosphere itself below –50°C (–58°F).

It wasn't the first seed to be sprouted in space, as Arabidopsis had been grown before in a spacecraft – but it did prove we could germinate a seed on a celestial body. It's the first step in working out how to feed people who are living away from Mother Earth for an extended period of time.

Despite some serious challenges en route, in much lower gravity, on a planet that has no atmospheric oxygen, humankind showed what could be done. Or, perhaps the seed showed what could be done.

It's possible to germinate seeds on the Moon, or in rocky places such as outside my dairy, not because of some amazing human ingenuity, but because of the seed itself. As well as the energy reserves to get it started – the carbohydrate and fat locked inside its germ – each seed also has its own microbiome, its own microbial signature, an entire genetic code along with the seed's own DNA, that allows it to burst into life.

This should come as no surprise. Microbes are everywhere, including in and on us, so of course they're also in and on plant matter. As we saw in Chapter 6, microbes in food actually enter our bodies through our gut, are taken into our own gut microbiome, and can send signals to our brain.

When you eat an apple, you could ingest about 100 million bacteria. But 90 per cent of those are in the seeds,[1] which most of us discard. The seeds of other plants can have even more. While research into the biome of seeds is really only just beginning, scientists already know that a single seed can contain up to 9000 species of micro-organisms (including fungi and archaea, as well as bacteria). These can number up to a staggering 2 billion individual microbes. In a single seed![2]

Forget the complexity of travelling to the Moon – a seed is more complex than we yet understand.

A seed's microbes exist not just on the outside, but also, vitally, within the very seed itself – a storehouse of bacteria, fungi and archaea that can help colonise new ground. These are, in large part, soil microbes, held in store so the seed has a ready collection of them for when it starts growing.

Microbes can enter the seeds maternally on the seed's outer coat. Paternally, many microbes can arrive through grains of pollen, or through the stigma. Much comes from the soil. And part of the microbiota can arrive from the guts of animals and reside on the outside of the seed.

In effect, the seed holds a library of many of the microbes that it will need to grow, allowing it to colonise dirt, not just soil. It's not a one-way street, though, and this could be where my barley starts to falter, as soil also provides vital microbes to the plant. Seeds sown in sterilised soil can survive, and some even come close to thriving, but the microbiome in the resulting plants is a shadow of what it can be if planted in healthy soil.

My barley struggles because the rock dries out very quickly. It struggles to get a foothold because the stone is easily dislodged. And it probably struggles because the crushed rock is only part of what the plant needs. The biological component is completely missing, and the microbial community that the seeds provide struggles to expand from the roots.

Microbes from soil, from the parent plant, from animals, are enshrined in and on each seed, awaiting potential germination. Again, we see that soil and plants and animals are intertwined, at the microscopic level. You can't affect one without affecting the other.

And we know that artificial fertilisers and pesticides interrupt the biome of living plants, and that those plants pass on their microbes to their progeny, through seeds. In the same way that antibiotic use in women has led to a decreased microbiome in their children over succeeding generations,[3] so too the use of pesticides and artificial fertiliser has decreased the microbial diversity in plants.

Add this to the reduced diversity and activity that has occurred thanks to the domestication of plants and cultivar selection noted in previous chapters, and it's clear that we're going in the wrong direction, overall.

It's all rather exciting when we hear that scientists have germinated a seed on the Moon. But all rather mundane when a gardener germinates a seed in their greenhouse, or in their plot.

And this is the problem. We, as a species, are thrilled and entranced by the new and the possible, by the mystique of space and the potential to grow food on Mars. We gaze up at the night sky and wonder, 'What if?' Very few have the same inclination to look down and ask the same question. In doing this, we neglect the miracle at our feet. At the same time as a Mars Rover searches for signs of life in the dirt of the red planet, the actual life we have on Earth is being squandered.

While the Mars explorer Curiosity was scouring the red planet's dusty surface in 2019, back on Earth, 245 million kilometres (152 million miles) away, the Australian state of New South Wales experienced 8000 hours of dust storms, losing several million tonnes of topsoil. Both events hit the news, but one received way more fanfare than the other. One was filled with hope; the other framed as hopeless, a natural consequence of drought – when in reality the erosion and loss was exacerbated by human damage to soil.

For many, the thought of Planet B is tantalisingly alluring, when, actually, the best we can do is keep a cotton seed alive for a bit over a week. Meanwhile, back on Earth, we're busy creating moonscapes.

What we still don't know about a plant's interactions with this planet's soil could challenge the world's best minds for decades. The microbiome of seeds was unknown until recently – when it was suddenly found that, of course, microbes are everywhere, controlling everything, including seeds, and the ground the seeds eventually grow in. Yet still we coat much of the 3.5 million tonnes of seed sold globally with fungicides,[4] things designed to kill up to half the microbiological diversity the seed was born with.

A seed is pre-charged with an invisible powerhouse, ready to burst forth and colonise soil. Or even colonise dirt, if conditions are right. It is armed with all the genetic code to be able to harness sunlight, and draw down carbon, feeding and being fed by a subterranean ecosystem. The germination of a seed is a powerful sign that a plant's photosynthetic cells will turn sunlight into sugar, to feed the earth, to feed wildlife, to feed livestock, to feed us. But it can only do so if we value the soil we have.

Of course, technology is news – especially in an era when so many are removed from the soil that gifts us life. We forget that technology has also brought problems, from the plough to pesticides to the Green Revolution. The growing of food has turned out to be destructive in so many ways, from habitat loss, to the intensive farming of animals, to poisoned waterways.

Now technology has taken a radical new approach to food creation, which doesn't require the use of farms at all. In recent years, it's been shown that by using electricity, along with the gases in the atmosphere, and certain microbes, you can produce nutrients – circumventing the need for plants. Or soil.

Companies such as Solar Foods are producing protein out of thin air, using electricity as the energy input. In fact, Solar Foods' stated aim is to separate food production from agriculture, freeing us from the drudgery and burden of having to farm, and from the potential to ruin more soil. In theory, we can use electricity to separate hydrogen from water, combine it with nitrogen from the air, using microbes in a fermentation chamber, and produce protein.

The result is Solein, marketed as 'the purest protein in the world'. Using a group of NASA-discovered microbes called hydrogeno-trophs in a fermentation process, it's fired with 'renewable' electricity to make a protein molecule from the atmosphere. The protein is, essentially, dead microbes. The inventors claim that this protein is neutral in taste and appearance, so it can be added to any food products you like.

As Solar Foods' website boasted at the time of writing, 'By offering a platform that disconnects food production from agriculture, Solein is protein as pure as it is sustainable. As if by magic.'[5] Hydrogenotrophs can convert carbon dioxide and hydrogen to protein, and release the greenhouse gas methane. Solar Foods feeds the bacteria nitrogen, in the form of ammonia.

Of course, the original magic was photosynthesis. And here we have humans again trying to outsmart it with food that is born in a factory. Food that is, according to the processors, 'pure'. Food that is proud to disassociate itself from farming, and flavour. It is simplified nutrition, produced by people with something to sell. Hydrogenotrophs, in theory, are more efficient at producing 'food' (carbohydrate and protein) than plants are – though they do need a lot of inputs in terms of buildings, fermentation chambers, electricity and the like. And I do wonder about the lack of flavour; protein so pure that it ignores the 'nutritional dark matter', the other 70,000 or so chemical compounds that we have evolved to taste and eat.

When Solein launched, it made massive global headlines, in part thanks to the vegan agenda of people such as the British journalist and activist George Monbiot. Protein from air means we won't need to farm animals. A lot of their website is given over to how much media they've attracted, and none to communities, societies and the actual complexities of nutrition.

Air Protein is another brand that is using a fermentation process to produce protein from the atmosphere that they describe as 'similar' to that found in meat – focusing on the new ultra-processed meatless meat market, along with fortifying cereals and pasta. NovoNutrients, another California-based start-up, is doing a similar thing. Unlike Solar Foods, their focus seems to be on producing animal feed (perhaps removing the need for wild fish meal to be used in fish farming, or the use of soy in factory farms).

It seems that there's an awful lot of money and time, and seriously big brains, working out ways to 'disrupt' the food scene. We're always on the lookout for the new, the shiniest, the interesting, the different.

More money, more time and more brains could be exercised in that thing that has already sustained us for all of human history – the living reality that a seed can grow into a plant that reaches its optimum nutrient density in healthy soil. The truth is that sunlight can be turned into food, with little or no energy inputs from us, and this – for all time – has been the basis of complex life on Earth. If only we had the humility to know it was so.

As a species we can, if we want, go down the path of Solein and other factory-made nutrients. They certainly grab the public's attention. They fire up the imagination of those bored by staid images of farming. But there are many great growers who don't fear hard labour on the land, who don't mind dirtying their hands harvesting the lettuce or turning the compost. There are those for whom the noble act of capturing sunlight and turning it into tucker for the rest of us to eat isn't a step back to primitive times, but rather provides a source of complex nutrition, ecological repair and meaningful employment. We know the further we get from soil and its beneficial microbes, the more ill we become as a population.

Farm-free food hits the headlines because it's a thing of science fiction, and because of the exciting leaps in technology it suggests. It's like watching Matt Damon grow a meal for himself on Mars in Ridley Scott's 2015 movie *The Martian*. The reality, however, is that the human race won't be moving to Mars any time soon.

This obsession with space seems ridiculous to soil scientists. Dr Kris Nichols, a soil microbiologist who worked with Sara F. Wright on glomalin, puts it really well. 'Our issue with colonising the Moon isn't that we don't have the technology. We know how to [get the people there],' she says. The problem is that we don't even know how to grow food on Earth yet. 'We're basically making the best land on this planet into the Moon.'[6]

Earth's reality is less Jetsons and more Old MacDonald. Over the millennia it has been hardwired into our noses and our guts that the best nutrition for us is encapsulated in a variety of things grown in healthy soil and consumed at their peak. We can chase the dream of interstellar missions, of nutrients grown in sterile laboratories. We can believe the myth that no animal dies when we grow the peas for our Impossible Burger, that we are genetically programmed to get all we need from Air Protein meals. Or, we can rejoice in the innate beauty of a mixed farm. The wondrous complexity of a properly managed ecosystem that abounds with things we *can* eat, and that we are *designed* to eat, and that is good for the farmer and good for the farmed – good for both the land and the life it contains.

For much of the last hundred or so years, modern growing has focused on how to simplistically manage an impossibly complex set of ecosystems. It has focused on killing things – fungi, insects, weeds, pathogenic bacteria and parasitic nematodes – and not enough on letting things live. We're in danger of looking to outer space and farm-free food without actually understanding what we, and our planet, really need to thrive.

We have the choice to devour with pleasure the fruits of the land, things that look like real ingredients, grown in soil that we care for as our nurturer. Or we can try to cleanse our lives of dirt and outwit nature and face the inevitable health and environmental consequences that we already know will arise.

Separating us from soil has already led to problems. Separating our food from soil will bring more of the same. The only way to heal the world and to live our best lives is to care for the earth that supports us. But can we do that with a population that's forecast to hit 10 billion before too long?

Well, yes. Of course we can. I'm about to tell you how.

CHAPTER 24

Loaves and Fishes: Feeding a Hungry World

When we talk about soil, we must talk about food. Nourishment is, after all, one of the main functions that soil does for us, and one we are intimately tied to. While some of the best soil lies under ancient forests, and some of the majesty of soil is best seen in the world's wild places, overall, only a tiny fraction of land is left alone. Much is managed woodland. And much is farmland. We grow food and farm animals on about half of the Earth's suitable land, suitable because it's the bit with topsoil. When we grow food, we impact soil – and how we treat soil will determine how we feed ourselves into the future.

As I write, the world population sits at about 7.8 billion. It's forecast to rise to about 9.6 billion by 2050[1] – meaning another 2 billion souls will arrive in the next 30 years or so.

To put this in context, in the year 1500, there were only about half a billion people in the world. By 1800 there were 1 billion people, a doubling in 300 years. It doubled again in 127 years; and again after only 47 years.

Despite a lot of talk about trying to control population growth, it's never been easy, and there's precious little we can do about it in a hurry. China's One Child policy, for instance, was abandoned due to the severe societal pressures it caused.

The thing is, the birth rate actually is declining. While the world population doubled in my lifetime since about 1970, it will take another 200 years to double again from 2020. Much of the population growth now will be from keeping people who have already been born alive for longer, rather than a fast birth rate.

Whatever way you look at it, we're in new territory. We have an awfully large number of souls to feed now, and even more in the future.

Can we do it? Well, it might pay to see how we have done so far.

It has been estimated that as we transitioned from foragers to settled agriculture, the amount of land needed to feed a person decreased from 10,000 hectares per individual, to only one hectare[2] – meaning farming was about 10,000 times more efficient than using more wild systems.

That was before the use of fossil fuels. Some say we now only need a quarter of a hectare (about half an acre) to feed each person.[3]

We've become really efficient at growing more food, and growing more food on the same land. That's why the global population has been able to explode exponentially. Of course, we eat very differently now from forager–hunter times. Less variety, more calories, for a start.

So what does the current global situation look like, nutritionally?

Out of 7.8 billion people currently alive, a full 3.7 billion people are malnourished.[4] This is despite the fact that we grow enough food for about 11 billion people, at least in macronutrients. As we know, the problem up until now hasn't been in growing food as such, but largely geopolitical, and in the unfair way in which food is distributed.

What's more, of the food we do produce, we waste about 40 per cent[5] – thrown out on farms, at food processors or wholesalers, as well as in homes. Some say closer to 50 or 60 per cent is lost to waste.[6]

So, cutting waste would feed more than the 10 billion we're expecting by mid-century. As would the political will to distribute food evenly.

Some argue that because 71 per cent of the world is water, it's a relatively untapped resource in terms of area that could step into the void. Can the oceans feed us? They're big, soak up a lot of sunshine, and feature strongly in our cultural imagination. Seaweed? Wild-caught seafood? Aquaculture? Well, only about 2 per cent of our calories come from the oceans, according to the UN's Food and Agriculture Organization,[7] so unless something very drastic or innovative happens, the seas aren't likely to take up the slack.

Even if we cut waste and reach political utopia, we'll have to rethink how we grow food, because our cropping land is losing fertility, and much grazing land is becoming desertified.

We spray 3 million tonnes of pesticide a year on crops,[8] and still lose about half of those crops to insects and disease. According to David Pimentel, the late professor of insect ecology and agricultural sciences at Cornell University, total grain production per capita has been declining since 1984; arable (cropping) land per capita has been declining since 1948; fish production per capita has been declining since 1980; and loss of food to pests has not decreased below 52 per cent since 1990.[9]

Even nitrogen fertiliser isn't saving us. According to Dr Kris Nichols, a soil scientist who's worked with the U.S. Department of Agriculture and the Rodale Institute, nitrogen efficiency is going down, not up. 'It takes more nitrogen fertiliser to grow a bushel of corn today than it did in 1960,' she reckons.[10]

In other words, we're losing ground. A lot of the problem, of course, is because we've been losing the *actual* ground. It's already costing us, in more ways than one. In the United States alone, corn farmers are already spending half a billion dollars more on nitrogen fertiliser every year because of lost soil fertility.[11]

Despite the Green Revolution, despite the fact that we have discovered ways to make plants grow bigger, quicker, and can magic up nitrogen fertiliser from thin air, the food system is broken. It's broken because we don't look at soil in the right way.

While Maria Helena Semedo at the Food and Agriculture Organization estimates we only have about 60 harvests left on much of our soil,[12] the first ever global soil survey in 2020 painted a more positive picture.[13] Yes, some land is impoverished and we are losing topsoil at tragic rates, with about a third of conventionally managed land having only a century or two of soil left; in fact, one in every six hectares of farmland will be eroded of topsoil in a century.[14]

The survey did show that a tiny fraction of soils in the hands of those practising conservation-style farming might last 10,000 years.[15] And there are a precious few places where humans are actually replenishing soils. We now know what it will take to save and restore soil.

What it does show is that it is time to get serious about how we're treating the ground that feeds us.

Soil isn't inert. It isn't something to mine. When we think of farmland, we need to consider it as part of the national estate, a resource held in private hands for public good. Not just our good in the short term – a period like a decade – but in the long term, something more like a millennium.

What happens in soil matters today, but it also matters for all of the tomorrows.

The Green Revolution, as we saw, did increase some yields, but most of the gain went into feeding ethanol plants and factory farms, not the poor in the world. When it did feed them, it altered soil structure for the worse. We lost soil *and* nutrient density of the crops. We farm more land, not less, despite the promise. By the time we have sprayed nearly half the world's arable land with glyphosate, and half is ploughed at least once a year,[16] soil is left gasping for breath.

So back to my original question: is there a way to feed 10 billion people and not bugger up the planet?

Yes, of course there is. But it will take a reimagining of earth, and a reframing of soil.

We've seen that Dark Earths, those of Africa and the Amazon, can stay fertile for centuries, or even millennia, without needing an armoury of chemicals.

We've seen that plants flourish in polycultures, where bacteria reach quorum sensing and start to thrive and nurture each other, as well as the plants that they nourish.

We've been witness to modern soil-building techniques such as Niels Olsen's Soilkee, which previous generations would have thought unimaginable.

We also know that we can do so much better with the resources we have right in front of us, not just within the structure and life of soil, but also in the food waste, the human waste, the dead.

We know that the more plants we can grow, drawing down carbon from the atmosphere and perhaps stabilising the climate a little, the more vibrant the animal ecosystem will be. A more vibrant ecosystem that encourages animals, wild and domesticated, to also help cycle nutrients that nourish soil.

We know that soil can be the genesis of rain – but that rain, especially inland, has to form from the moisture given off by healthy plants, grassland, shrubs and trees that ring our continent and other large land masses.

We know that grazing land is better at storing carbon – and also rates better on most other measures of soil health and ecosystem diversity – than cropping land, and that we can't eat grass, but a ruminant can.

But we also know that the whole globe can't all eat over 100 kilograms (220 pounds) of meat a year, as we do in Australia and the United States,[17] without something collapsing – be it soil, or climate, or population. Or a combination of the three.

For the last 100-plus years, we've reduced soil to the chemical and physical, and added the nutrients we think it needs.

It's time we celebrated the biological, and restored land in quantities that can help slow climate change. We need to consider soil as the massive store of gases, life and nourishment in the holistic sense that it is: soil as an almost incomprehensively complex underground super-organism, capable of sustaining life on Earth. It can feed 10 billion, easily, if we do that.

I don't believe in God. But I do think there's something to be said for the countless centuries where we considered ourselves at the mercy of soil, at the whim of the gods of harvest: Demeter, Ceres, Tudi Gong, Heqet. They at least gave us humility when we dealt with earth.

Science has shown us ways to improve crops. And that's been wonderful at times. It has delivered an arsenal of chemicals, similar to the ones we use in human health, which can help avert catastrophe. Technology has allowed us to remove the back-breaking work from growing food.

We need to look at breeding variety again to regenerate soil and farming, to embellish cooking, and focus more on nutrient density. We need to value biodiversity. Wouldn't it be marvellous if all the world's growers could think of growing today's food as nurturing the soil ready for next year's harvest? Most of what we have done to soil is through ignorance, or hubris, believing we've got it all worked out, when in reality we haven't.

So I have time for the woo-woo. For the worship of the unknown. Call them soil gods if you will, those invisible billions; underground microbes, harnessing the best of the earth for the plants we eat. Call them deities if you want, those 70,000 or so chemical components in food, and their as yet un-numbered secondary metabolites. They defy our powers of logic, and, like compost, ask for faith in a system of life begetting life, despite the inability of reductive science to quantify it.

If science has shown us anything, it's that mother nature is on our side. We just have to work with her, not battle her constantly.

I have learned to have an almost transcendental appreciation of the complexity of soil biology, because it isn't all knowable now. We can use science to our advantage – but reductive science has time and again let us down when it comes to understanding the utter complexity of life on Earth, of life *in* earth, and how our diet improves our life.

The good news is that a movement is underway to treat soil like the biological organism it is. Growers will embrace soil health more quickly if we buy from those who put it at the heart of everything they do. There's no single right way; soil is agnostic. Good farming can take many guises, including the occasional and judicious use of chemicals that I wouldn't otherwise consider right for my land.

There's no need to wait for change to happen. At home, we can believe in soil right now. In its ability to harness life – the kind of nutrient-dense life our bodies are designed to devour. We can nurture our own plot, swap with our neighbours, buy off a local grower where we can. We have the power to support those who support soil. And we can take life into our own dirty hands. We've seen that urban gardens can easily feed large swathes of the population, providing meaning, sustenance, joy and beauty all in one hit. Urban gardens could well be our saviour again.

Of course, the future depends on how you view soil. I once saw it as something to rinse off the spuds. But now I view it as a living, breathing, seething mass. Whenever I see a multi-species crop, or a forest, or a verdant garden, I know that they are just the visible signs of a thriving ecosystem underground. Bare soil breaks my heart. Watching a plough carve through a paddock feels like a knife passing through flesh, the wound obvious, the scar not quick to heal. Driving past paddocks brown with herbicide, I can almost feel the subterranean microbes gasping, like a constriction in my chest. I imagine the carbon blowing off into the air.

These days, when I walk in a forest, I visualise countless worms devouring organic matter at the cusp where soil and fallen leaves meet. I can imagine the worms' bodies moving 'sleeping beauty' bacteria from place to place, their poo enriching soil, creating humates and helping to break down rock. I imagine the thousands of metres of fungal hyphae that link each living tree to others, as the plants use the network to share stories of attack, and life, and death.

When I walk the land of regenerative farmers, I not only hear the birds and frogs, but also think of the arthropods, the spiders, insects and crustaceans, that help hold up each step I take. I imagine the outline of the earth in nematodes. I see a crop in a field and visualise all the bacteria, archaea, and protists cycling nutrients, ensuring that nitrogen is available in the rhizospheres. I ponder the fungi that link roots, like invisible capillaries, seeking out minerals and feeding the plants life.

If we think we can outsmart nature, using monocultures and a chemical cocktail of fertilisers and pesticides, we're wrong. We've tried it before, and lost. If we lose this time, the consequences could be catastrophic. We have to be ready to feed 10 billion souls, and counting, in less than three decades.

It's doable, using the technologies and mindset I've outlined in this book. It's doable and it's urgent.

Soil is invisible, much of the time. It doesn't rate as sexy. Most people don't see it as vital, despite its role in protecting us, sustaining us, and healing us. Soil disappeared from our consciousness as we separated ourselves from the land, from the hand-reared nature of our food. As we moved from growing to just consuming, we took our minds off what the subterranean ecosystem needs to thrive. What it needs to be its best self. Worse than that, we've continually treated soil like dirt. And soil won't thank us for the slight.

If we don't tend the land with a mind to all the life it contains, we risk turning more soil to dirt. And that could well come back to bite us.

It's time to re-envision soil. To hold it in high esteem, not treat it with disdain. Instead of debasing it, degrading it, or even ignoring it, we need to feed the soil, nourish it where we can, in all the ways open to us as individuals, as communities, as humanity. And then soil will continue to nourish, nurture and sustain us, for all the tomorrows we hope to have. We just need to re-imagine the world, from the ground up.

Soil FAQs

What is soil?

Soil is mostly crushed-up rock (crystals of sand, silt and clay), with organic matter and living microbes. It also has an association with a living plant's roots, or has recently had such an association. About 95 per cent of soil mass is crushed rock (minerals), about 5 per cent is organic matter (things that have been alive at some point), and about 0.5–1 per cent is living microbes. Up to 25 per cent of living soil is air, by volume, and that air is vital for healthy soil.

How is soil made?

Lichens, fungi, worms and microbes can help break down rock, but most comes from other actions such as freezing and thawing, glaciers or gradual erosion. Organic matter comes from dead leaf matter and animals – so, dead roots, leaves and the like, along with insects, microbes and so on. Plants feed the underground ecosystem with sugars and other organic compounds, essentially pushing organic matter into the soil, because they can construct carbohydrate out of thin air, thanks to photosynthesis. Nature makes soil at a similar rate to it being lost, but humans have typically degraded soil faster than it is made.

What's the best thing I can do for my soil?
Water it, if you're in a place where it can dry out. That's the most important thing in the first instance. Get something, anything, to grow, because that starts the biology happening, and starts to store carbon. The next best thing you can do is to add carbon – so, mulch it, and don't leave any bare earth. Adding compost is the best carbon, because it is pre-digested, ready for the microbes to use, and it already has glycoproteins in it that give soil better structure. Compost also contains chemicals that enhance plant growth and help nurture the microbes in your soil.

Adding biochar to soil is rarely going to hurt, either, though many suggest soaking it in compost tea first to inoculate it with microbes.

Should I get a soil test done?
Soil tests can tell you a bunch of things about your soil, like its pH (a measure of acidity), and the presence of minerals and the like. But soil tests don't give you a measure of exactly how available nutrients are to your plants, because the microbes living in the soil are actually what make the nutrients bioavailable (in other words, available in a form that the plant can take up). Use soil tests as a guide, especially for minerals such as boron and manganese that may be absent, or in too small an amount, in your soil. But generally, home gardeners can manage soil life and nutrient availability without needing to resort to expensive tests.

I've seen that you can actually buy microbes to add to your soil. Are they a con, or do they work?
Adding living – and they *must* be living – microbes to the soil can be helpful. But more importantly, because microbes are airborne, and often might already exist in small numbers in your soil, you have to create the conditions for them to thrive. Compost, broken-down manure and mulch can all create food for the microbes to rebound in numbers. Get that right, and any microbes you already have, or do add, will also be able to thrive.

You keep banging on about how great soil is, but I've heard of diseases in soil that will attack my tomatoes or spuds if I don't rotate the crops. Aren't you just focusing on the positives?

Nature doesn't grow a monoculture, but humans do, even in our small home gardens. Pest species certainly do exist, and can be managed by rotating crops, or planting multi-species beds. You can also 'fumigate' your soil by planting things such as mustard leaf, which act as a nice biosecurity option, while still keeping living roots in soil. Not all soil life is wonderful (like botulism and legionnaires' disease), but the vast, vast majority is benign or helpful. Aim for harmony, and to work within nature, not against it. Growing food in a functioning ecosystem is the goal. As growers, we manage the ecosystem to some extent, because it's different to an entirely wild or native ecosystem.

References & Recommended Reading

This book cites hundreds of sources, many of them scientific papers that are available online. For the full set of footnoted references, Murdoch Books invites readers to visit our online portal, which takes you directly to the relevant study, article, interview or book for each entry. Visit murdochbooks.com.au, search for 'Soil' by Matthew Evans and click on the References tab. Selected books the author found a valuable source of information are listed below.

Andrews, Peter, *Back From the Brink: How Australia's Landscape Can Be Saved*, ABC Books, 2006

Balfour, Eve, *The Living Soil*, Soil Association, 2006 [1943]

Brown, Daniel G. (editor), *Land Use and the Carbon Cycle: Advances in Integrated Science*, Cambridge University Press, 2013

Darwin, Charles, *The Formation of Vegetable Mould Through the Action of Worms*, John Murray, 1881

Evans, Matthew, *On Eating Meat*, Murdoch Books, 2019

Evans, Matthew, *The Real Food Companion*, Murdoch Books, 2010

Howard, Albert and Wad, Yeshwant D., *The Waste Products of Agriculture*, Oxford City Press, 2011 [1931]

Ingham, Elain, *The Foodweb in Compost*, Soil Foodweb (self-published), 2009

King, F. H., *Farmers of Forty Centuries*, Dover, 2004 [1911]

Marshall, Tim, *Composting: The Ultimate Organic Guide to Recycling Your Garden*, ABC Books, 2008

Massy, Charles, *Call of the Reed Warbler*, University of Queensland Press, 2017

Masters, Nicole, *For the Love of Soil*, Printable Reality (self-published), 2019

McCoy, Peter, *Radical Mycology*, Chthaeus Press, 2016

Montgomery, David R., *Dirt: The Erosion of Civilisations*, University of California Press, 2012

Montgomery, David R., *Growing a Revolution*, Norton, 2017

Montgomery, David R. and Biklé, Anne, *The Hidden Half of Nature*, Norton, 2016

Palmer, Nigel, *The Regenerative Grower's Guide to Garden Amendments*, Chelsea Green Publishing, 2020

Pascoe, Bruce, *Dark Emu*, Magabala Books, 2014

Provenza, Fred, *Nourishment*, Chelsea Green Publishing, 2018

Raubenheimer, David, and Simpson, Stephen J., *Eat Like the Animals*, HarperCollins, 2020

Savory, Allan, with Butterfield, Jody, *Holistic Management: A Common-sense Revolution to Restore Our Environment*, Island Press, 2017

Sheldrake, Merlin, *Entangled Life*, Bodley Head, 2020

Solomon, Steve, *The Intelligent Gardener: Growing Nutrient Dense Food*, New Society Publishers, 2012

Sombroek, Wim, *Amazon Soils*, Wageningen University, 1966

Stamets, Paul, *Mycelium Running*, Potter, 2015

Stika, Jon, *A Soil Owner's Manual*, Createspace Independent (self-published), 2016

Tickell, Josh, *Kiss the Ground*, Enliven Atria, 2017

Tree, Isabella, *Wilding*, Pan MacMillan, 2019

Williams, Paul R., *Feast on Phytochemicals*, Vegetation Management Science (self-published), 2019

Yeatman, Marwood, *The Last Food of England*, Ebury Press, 2007

Yong, Ed, *I Contain Multitudes*, Vintage, 2016

Acknowledgements

I am indebted to a whole range of delightful soil nerds, from the farmers and gardeners to the microbiologists and researchers; from ancient cultures to the most recent scientific findings. What you have done to bring this majestic story of soil into view is nothing short of remarkable.

A book like this takes form over many years, and is beaten into shape by the best of publishers. A massive thanks must go to Jane Morrow, the best of the best, from Murdoch Books, who read an early draft that was twice the length of the final product. She could see the soil for the dirt. I don't have words to express how grateful I am that I was able to hand you a wordy, worthy set of still-forming ideas and trust you could nurture the book within.

To Katri Hilden, whose edits are thorough, searching, thought-provoking and precise, I really do appreciate you making me sound like a better writer than I am. Gratitude also to the eagle-eyed proof-reader, Dannielle Viera, who knows better than me the distance to Mars. And to Justin Wolfers and the rest of the team at Murdoch, thanks for the ongoing and unstinting support over the years, and particularly for helping bring an earthy idea to life. Who knew soil could be so fascinating?

This book was written at a time of great global and personal challenge, with COVID-19 plaguing the world, and farming injuries limiting my

263

ability to care for soil. So it is only through the great work of the team on our farm that I was able to dedicate the time, and resources, to writing it.

Rob Cartledge carried a heavy load, including buckets of pig food (and running the commercial kitchen), while listening to stories of my wonder for soil. Rowan Jordan cared for the livestock, brush-cut the weeds and was regaled with every new fact on subterranean life I could find. To the whole team, including Jo Duffy and Jane Herring behind the scenes, Juan David Ramirez Velasquez, Georgie Moon, Maria DeBraganca, Tom Kadwell and Merlin Glendining in the kitchen, and Zara Trihey, Junaidi Susantio and Catherine Arsaut on the floor. I know I was too often absent, but I do hope you like the results of all our labour. And to the mob who tend the soil in our market garden, the part that gives us most variety on our land: Sadie Chrestman, Jess Knight, Clare Aston, Gemma Sherlock and Nadia Danti, you make the farm magnificent.

Thanks to all those who read and commented on the book, including Kirsten Bradley, Costa Georgiadis, Matt Tack, Valerie Matsumoto, Isabella Tree, Nadia Danti, Gabrielle Chan, Simon Thomsen, Tim Thatcher, Rebecca Huntley, Damon Gameau, Steve Oliver, Charles Massy, Peter Gilmore, Alexx Stuart and others; a HUGE soil-covered hug of thanks. I'm supremely grateful to Niels Olsen and all the other farmers and gardeners from Ballarat to the Margaret River for site visits and inspiration. Thanks to Jack Pascoe, who helped demystify the role and impact of Aboriginal traditional cool burns in Australia's landscape, and why they are so poorly researched. And it's a massive debt of gratitude to soil scientist Dr Kris Nichols, who took time out from her work in the United States to elucidate glomalin and a whole bunch of other really cool soil-related concepts.

To Sadie, my partner in life and soil, who edited, researched, critiqued, cooked, encouraged, and ran the farm in my almost complete absence while I barricaded myself in the office, trying to meet deadlines. I owe you more than just an electric mower next Christmas. Perhaps something to direct-drill and improve carbon in the paddocks?

Above all else, this book is an ode to our son Hedley, because feeding soil is all about feeding the next generation. And the one after that, and the one after that.

Index